73

Fortschritte der Chemie organischer Naturstoffe

Progress in the Chemistry of Organic Natural Products

Founded by
L. Zechmeister

Edited by
W. Herz, G. W. Kirby,
R. E. Moore, W. Steglich,
and Ch. Tamm

Authors:
T. Fukai, T. Nomura

Springer-Verlag Wien GmbH

Prof. W. Herz, Department of Chemistry,
The Florida State University, Tallahassee, Florida, U.S.A.

Prof. G. W. Kirby, Chemistry Department,
The University of Glasgow, Glasgow, Scotland

Prof. R. E. Moore, Department of Chemistry,
University of Hawaii at Manoa, Honolulu, Hawaii, U.S.A.

Prof. Dr. W. Steglich, Institut für Organische Chemie der Universität München,
München, Federal Republic of Germany

Prof. Dr. Ch. Tamm, Institut für Organische Chemie der Universität Basel,
Basel, Switzerland

Originally published by Springer-Verlag/Wien in 1998

Library of Congress Catalog Card Number AC 39-1015

Typesetting: Thomson Press (India) Ltd., New Delhi

Graphic design: Ecke Bonk
Printed on acid-free and chlorine-free bleached paper
SPIN: 10634021

With 30 Figures

ISSN 0071-7886
ISBN 978-3-7091-7333-6 ISBN 978-3-7091-6480-8 (eBook)
DOI 10.1007/978-3-7091-6480-8

Contents

List of Contributors

FUKAI, Prof. Dr. T., Faculty of Pharmaceutical Sciences, Toho University, 2-2-1 Miyama, Funabashi, Chiba 274, Japan

NOMURA, Prof. Dr. T., Faculty of Pharmaceutical Sciences, Toho University, 2-2-1 Miyama, Funabashi, Chiba 274, Japan

Phenolic Constituents of Licorice (*Glycyrrhiza* Species)

T. NOMURA and T. FUKAI
Faculty of Pharmaceutical Sciences, Toho University, Miyama, Funabashi, Chiba, Japan

Contents

1. Introduction

The Leguminosae is an economically important family in the Dicotyledonae with many cultivated species, *e.g.*, beans and peas. The family also contains many well-known medicinal plants. It is composed of 17,000 or more species that constitute nearly one twelfth of the world's flowering plants (*1*). Traditionally the family has been divided into three subfamilies, Caesalpinioideae, Mimosoideae and Papilionoideae, which are sometimes recognized as separate families Caesalpiniaceae, Mimosaceae and Papilionaceae. The International Code of Botanical Nomenclature permits alternative nomenclatures, the family names being replaced by Fabaceae, Fabales and Faboideae, and this usage will be common (*2*).

Licorice (liquorice, kanzoh in Japanese, gancao in Chinese) is the name applied to the roots and stolons of some *Glycyrrhiza* species (Fabaceae) and has been used by human beings for at least 4000 years. The earliest written reference to the use of licorice is contained in the Codex Hammurabi dating from 2100 B.C., and the subsequent history in the West has been described in the earlier reviews (*3–6*). In the Far East, references to the effectiveness of licorice are contained in the "Shen Nong Ben Cao Jing", the first Chinese dispensatory whose original anonymous volumes probably appeared by the end of the third century (*7*, *8*). In the Chinese book, 365 crude drugs are classified into three classes (upper: plants with lowest side-effects and nontoxic useful for health care; middle: plants that are nontoxic or possess only weak toxicity in whose use care must be exercised; lower: toxic and only for clinical use). Licorice is described as belonging to the upper class and is recommended for its life-enhancing properties, for improving health, for cures for injury or swelling and for its detoxification effect (*9*). One hundred ten prescriptions are recorded in the earlier Chinese medicinal book "Shang Han Lun" as of C.E. 1065 (Its original version "Shang Han Zu Bing Lun" probably appeared between the third and fifth century), where seventy prescriptions include licorice (*8*, *10*).

Following extraction, the herb yields the licorice products of commerce which are used as flavoring for American type tobaccos, chewing gums, candies, *etc.* (*11–13*), sweetening agents (*11*, *14–16*), and as a depigmentation agent in cosmetic (*17–19*) and pharmaceutical products, *e.g.*, anti-ulcer drugs (Aspalon®, Caved-S®, *etc.*) (*20*, *21*). Clinical studies of licorice make it one of the most thoroughly studied herbs (*6*); these include its toxicity (*22*), antimutagenic activity (*23*), anti-ulcer effect (*24*), protective action for hepatotoxicity (*25*), antitumor promoting activity (*26*), anti-caries effect (*27*), antimicrobial effect (*28*),

References, pp. 110–140

etc. The crude drug is also the raw material for the drug glycyrrhizic acid (**1**) and its derivative carbenoxolone (**5**). But the most important industrial use of the herb is in the production of additives as flavor and sweetening agents (*13*, *29*).

The genus *Glycyrrhiza* consists of about 30 species (*30*) and chemical researches have so far been carried out on 13 of them (*31*). *G. glabra* L. is distributed in Africa, Asia, Australia, Europe, the Middle East and North America. The species includes some varieties frequently called Spanish or Italian licorice (*G. glabra* L. var. *typica* Reg. *et* Herd), Persian or Turkish licorice (var. *violacea* Boiss.), Russian licorice (var. *glandulifera* Reg. *et* Herd. or var. *glandulifera* Waldst. *et* Kit.) and European licorice (in China). *G. uralensis* Fischer grows in Asia and Russia (Far East) and is termed Ural, Mongolian or Chinese licorice, and *G. korshinskyi* Grigorj. is widely distributed in the former Soviet Union. *G. lepidota* Pursh is found in Central and North America and called wild licorice. *G. echinata* L. (sometimes called *G. macedonica* Boiss. *et* Orph.) occurs in Asia, Europe, the Middle East, Iran and European Russia. *G. aspera* Pallas. is a small plant, relatively unimportant in commerce, and is found in Asia, the Middle East, Europe and in Russian Asia (*31*). *G. triphylla* Fischer *et* Mey. grows in Afghanistan, Iran, Pakistan and Russian Asia, but differs from the other *Glycyrrhiza* species in morphology as well as chemistry. The species is now regarded as separate from *Glycyrrhiza* and may be transferred to the genus *Meristotropis* (*30*, *32*).

Many *Glycyrrhiza* species are found in China, but in the Northeast only *G. uralensis* and *G. pallidiflora* Maxim. (prickly fruit licorice, pseudo-licorice; inu kanzoh in Japanese) grow. On the other hand, in northwestern China *G. inflata* Batalin, *G. aspera*, *G. glabra*, *G. uralensis* and yellow licorice (*G. eurycarpa* P.C. Li = *G. uralensis* × *G. inflata*, a newly named species previously conflated with *G. korshinskyi*) are distributed (*33–35*). The pure species are rarely found in the region because of facile hybridization. *G. yunnanensis* P.C. Li (Malay licorice) is found in southwestern China and *G. squamulosa* Franch. grows in the central region (*33*, *36*). Annually some 3,000 – 6,000 tons of licorice are imported into Japan (*37*) where one fifth is used clinically in traditional Sino-medicine (Kanpo-yaku in Japanese). Chinese licorice (*G. uralensis*) is used for this purpose owing to its pure sweetness without bitter taste. However the supply from China is decreasing.

As for the components of *Glycyrrhiza* species, the essential oil (*11*, *38–40*), and the occurrence of alkaloids (*41–45*), polysaccharides (*46–62*), a polyamine (*63*), triterpenes (*64–67*), fatty acids (*68*) and amino acids (*69*) as well as triterpenoid saponins and flavonoids have

Table 1. *Review Articles of Licorice published between 1967–1995*

Author	Item	Reference numbers	(Period)[a]	Ref.
RUSSO	Sapogenin	23	(1937–1967)	*(70)*
OTSUKA	History, Chinese medicine	16	(–1969)	*(9)*
KUMAGAI	Pharmacology	31	(1946–1971)	*(71)*
WATANABE	Pharmacology	33	(1944–1971)	*(72)*
OURA *et al.*	Pharmacology	39	(1955–1972)	*(73)*
SHIBATA and SAITOH	Phenol, pharmacology	78	(1934–1972)	*(74)*
YAMAMOTO	Pharmacology, therapeutics	9	(1959–1972)	*(75)*
YANO	Pharmacology	41	(1950–1972)	*(76)*
BALTASSAT-MILLET *et al.*	Pharmacology	98	(1948–1975)	*(77)*
SHIBATA and SAITOH	Pharmacology, phenols	39	(1934–1978)	*(78)*
GIBSON	History, pharmacology	45	(1948–1979)	*(4)*
LIU	History, Xinjiang licorice	26	(1597–1979)	*(79)*
MAEDA	Sweetening	9	(1851–1979)	*(11)*
NISIMOTO and YASUDA	Production	10	(1970–1980)	*(80)*
SHIBATA	Sapogenin, phenol	26	(1935–1980)	*(81)*
LUTOMSKI	Saponin, therapeutics	101	—	*(82)*
JIANG *et al.*	Sweetening	18	(1967–1981)	*(16)*
SHIBATA	Phenol, sapogenin, pharmacology	15	(1959–1985)	*(83)*
CAI *et al.*	Saponin, sapogenin	71	(1935–1987)	*(84)*
LIU and LIU	Phenol	39	(1972–1988)	*(85)*
AYABE	Callus culture	52	(1975–1989)	*(86)*
BIELENBERG	Saponin, pharmacology	42	—	*(87)*
KIKUTI	Sweetening	14	(1979–1989)	*(88)*
FENWICK *et al.*	History, phenol, sapogenin, sweetening	120	(1943–1990)	*(5)*
LUTOMSKI *et al.*	Pharmacology	88	(1948–1990)	*(6)*
KITAGAWA	Saponin, phenol	6[b]	(1974–1991)	*(89)*
YONEDA	Production	7	(1974–1991)	*(29)*
HIAI and NAGASAWA	Pharmacology	60	(1962–1992)	*(90)*
KITAGAWA and HORI	Saponin, phenol	29	(1975–1992)	*(91)*
BAKER	Pharmacology	46	(1948–1992)	*(92)*
YIN and GUO	Whole items (hand book)	308	(1946–1992)	*(93)*
HU and SHEN	Therapeutics	36	(1958–1993)	*(94)*
JIA and QIU	Phenol (aerial parts)	20	(1966–1993)	*(95)*
SAITO	Molecular genetics	3[b]	(1990–1994)	*(96)*
FENG	Studies in China, phenol	9	(1981–1994)	*(97)*
ILDIS and CHCD	Structure (data base)	95	(1949–1994)	*(31)*

[a] The period means published years of collective citation references in the review articles.
[b] Numbers of references with respect to licorice. The article encloses the other topics.

been reported *(31)*. The earlier reviews of licorice are listed in Table 1. The present article briefly reviews the chemistry and biological activity of saponins obtained from *Glycyrrhiza* species and is devoted mainly to the phenolic constituents with isoprenoid substituent(s) such as the 3-methyl-2-butenyl (prenyl) group. New methods used for structure determination of prenylated phenols (flavonoids) found during in our chemical study of the phenolic compounds are also described.

2. Triterpenoid Saponins

2.1. Glycyrrhizic Acid

Glycyrrhizic acid is the major triterpenoid saponin in licorice root and is used frequently as a tool for recognizing the herb. This saponin has also been called glycyrrhizin, glycyrrhizinic acid or glycyrrhetinic acid glycoside and has been obtained from *G. glabra*, *G. uralensis*, *G. inflata*, *G. aspera*, *G. korshinskyi* and *G. eurycarpa* (*36*, *98–100*). The presence of glycyrrhizic acid has been known for one hundred years (*3*) and many investigators have studied the structures of the aglycone (**1**) and the saponin for its biological activity (*101–125*). These studies established its chemistry as that of an oleanane type triterpene. The structure of glycyrrhizic acid had been assigned as 3-*O*-[β-D-glucuronopyranosyl-(1 → 2)-α-D-glucuronopyranosyl]glycyrrhetic acid. In 1989, structure elucidation of the saponin was finished using modern NMR techniques and the structure was established as **2** (3-*O*-[β-D-glucuronopyranosyl-(1 → 2)-β-D-glucuronopyranosyl]-glycyrrhetic acid) by Khalilov *et al.* and the revised structure (**2**) was also confirmed by Shibata (*126*). This structure (**2**) had also been assigned to uralsaponin A obtained from *G. uralensis* and *G. eurycarpa* (*127*, *128*). Thus the structure elucidation of uralsaponin A needs further experiments.*

Glycyrrhizic acid (**2**) and its potassium and ammonium salts have an intensely sweet taste. This taste is completely removed by hydrolysis of the saponin to its aglycone, glycyrrhetic acid (**1**).† The distribution of saponins in different organs of *G. glabra* has been reported by Tabata *et al.* (*66*); glycyrrhizic acid is found in the thickening root and the stolon, but not in the seed, leaf and stem. On the other hand, soyasaponins are detected in the seed, hypocotyl and rootlet.

---

* The saponin could not be detected in a recent HPLC study of *Glycyrrhiza* species collected in China (*36*, *225*).

† The sapogenin is also called glycyrrhetinic acid, glycyrrhetin or enoxolone (International Nonproperty Name (INN) with WHO).

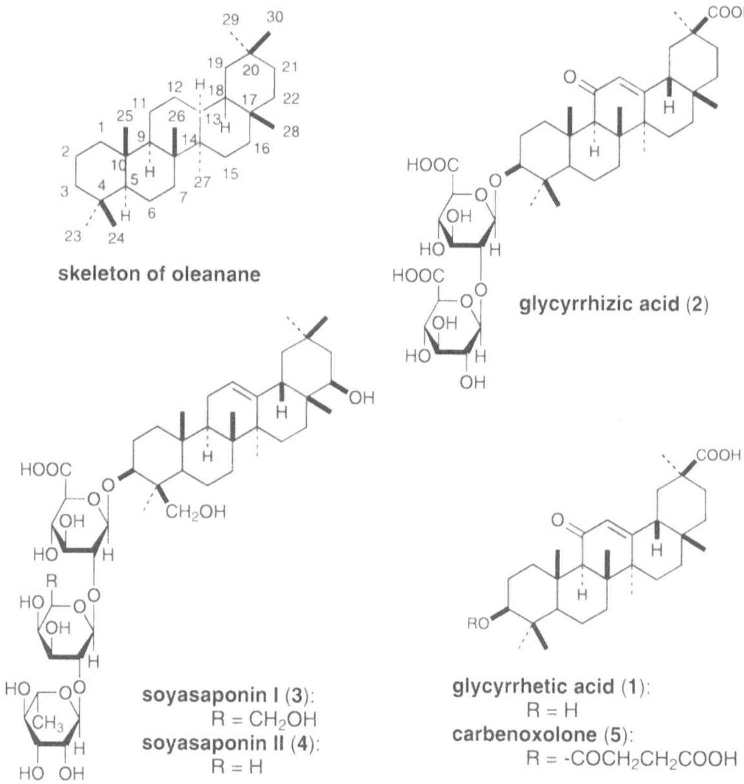

Fig. 1

Many investigators have studied the plant tissue culture of *Glycyrrhiza*, but the production of glycyrrhizic acid in this manner is only reported in the patent literature (*129–137*). The production of soyasaponins I (**3**) and II (**4**) in cell suspension cultures of *G. glabra* has been reported by TABATA *et al.*, and the contents are greatly influenced by plant growth hormones (*138*). The production of saponins by hairy root cultures transformed by Ri plasmids of *Agrobacterium* has been reported by SANKAWA *et al.* (*139*). One of the liquid culture lines of the hairy roots of *G. uralensis* produced glycyrrhizic acid in a yield of 4.7%.

Glycyrrhizic acid and structurally related saponins possess a sweet taste and flavor-potentiating characteristics and have been employed industrially for this reason (*15*, *140*). The sweet taste of glycyrrhizic acid (170 times as sweet as sucrose) is not felt immediately. Thus the reagent is usually used along with other sweetening agents such as sucrose, glucose, stevioside, *etc*. The saponin is also used to mask the salty aftertaste of soy sauce and salty foods (*141*). Some glycyrrhetic acid glycosides have been synthesized by MIZUTANI *et al.*, who examined their sweetness (*15*, *142*, *143*). Among them, glycyrrhetic acid monoglucuronide is sweeter than glycyrrhizic acid (about 5.5 times) and its flavor-potentiating action is more effective for a milky, chocolate and coffee flavor.

Glycyrrhizic acid (ammonium or potassium salt) and its derivative carbenoxolone (INN, 3-carboxy-1-oxoprepoxyglycyrrhetic acid, **5**) are used in the treatment of peptic ulcers and other gastro-intestinal disorders and as an antiinflammatory agent (*21*, *144*, *145*). These agents are metabolized by the same metabolic pathway, but differ in the time course (*90*). In treatment of chronic hepatitis and as a precaution against transfusion hepatitis, glycyrrhizic acid is also used by injection in large amounts (*146*, *147*). Japanese researchers have reported the inhibitory effect of glycyrrhizic acid on the infectivity and cytopathic activity of human immunodeficiency virus (HIV) (*148*). Clinical studies of the saponin conducted in Japan have shown promising results for HIV carriers (*149–157*). Intravenous injection of high doses of glycyrrhizic acid brings about improvement in liver function that is damaged in many HIV positive hemophiliacs. This therapy also prevents the side effects of azidothymidine (AZT). The researchers expect that the treatment protects asymptomatic HIV carriers from AIDS (*150*, *151*, *156*).

2.2. Structures of Minor Sapogenins

About fifty other sapogenins have been isolated from *Glycyrrhiza* species. Their structures are shown in Figs. 2–6.

2.2.1. Glycyrrhiza glabra

ELGAMAL *et al.* have isolated a minor sapogenin, liquoric acid (**6**), from *G. glabra*, and proposed a structure having an intramolecular ether linkage (*158*). Thirteen minor sapogenins, glabrolide (**7**), 11-deoxoglabrolide (**8**), liquiritic acid (**9**), isoglabrolide (**10**), 11-deoxoglycyrrhetinic acid (**11**), 3β-hydroxy-11,13(18)-oleanadien-30-oic acid (glypallidifloric acid, **12**), glycyrretol (glycyrrhetol)* (**13**), 21α-hydroxyisoglabrolide (**14**), 24-hydroxyglycyrrhetic acid (**15**), 24-hydroxy-11-deoxoglycyrrhetic acid (**16**), 18α-hydroxyglycyrrhetic acid (**17**), liquiridiolic acid (glyyunnansapogenin B_1, **18**) and 24-hydroxyliquiritic acid (**19**), have been isolated from this species by CANONICA and RUSSO *et al.* (*159–166*). Glabric acid has been isolated by BEATON and SPRING (*167*), and the structure (**20**) was elucidated by mass spectroscopic methods by ELGAMAL and FAYEZ (*168*). The isolation of 3,24-dihydroxy-11, 13(18)-oleanadien-30-oic acid (**21**) as a methyl ester has been reported by BOGATKINA *et al.* (*169*).† Uralenic acid has been isolated from *G. uralensis* and *G. glabra* by KIR'YALOV *et al.* (*170*, *171*), and the compound was identified as 18α-glycyrrhetic acid (**22**) by BELOUS *et al.* (*172*). The compound had been derived by isomerization of 18β-glycyrrhetic acid (**1**) under basic conditions (*124*). ELGAMAL and EL-TAWIL have isolated 28-hydroxyglycyrrhetic acid (**23**) from the same species (*173*).

The lactones **7**, **8**, **10** and **14** might be artifacts formed during the isolation procedures and the parent compounds of **10** and **14** have not yet been isolated. The parent compounds of **7** and **8** have been isolated as described in the following subsections.

2.2.2. *Glycyrrhiza uralensis*

ZHANG and her co-workers have reported the isolation of 24-hydroxyglabrolide (**24**), uralenolide (**25**) and glyuranolide (isolated as a methyl ester) (**26**) from *G. uralensis* (*174–177*). A formyl ester (**27**, Fig. 2) of glabrolide (**7**) and 22β-acetoxyglycyrrhetic acid (**28**),# the latter considered as the parent compound of glabrolide (**7**), have been isolated by SHEN *et al.* (*178*). The sapogenins **7**, **12–15**, **17** and **22** have been also isolated from this species (*174*, *175*, *179*).

* The trivial name "glycyrretol" was used in the English abstract of the original paper, and "glycyrrhetol" has been used in Chemical Abstracts (*163*).

† A glycoside of 3β,24-dihydroxy-11,13(18)-oleanadien-30-oic acid (licorice-saponin K2, **67**) has been isolated from *G. uralensis* (*205*).

The trivial name "22β-acetoxyglabric acid" was used for the compound (22β-acetoxy-3β-hydroxy-11-oxo-12-oleanan-30-oic acid) in the English abstract of the article (*178*).

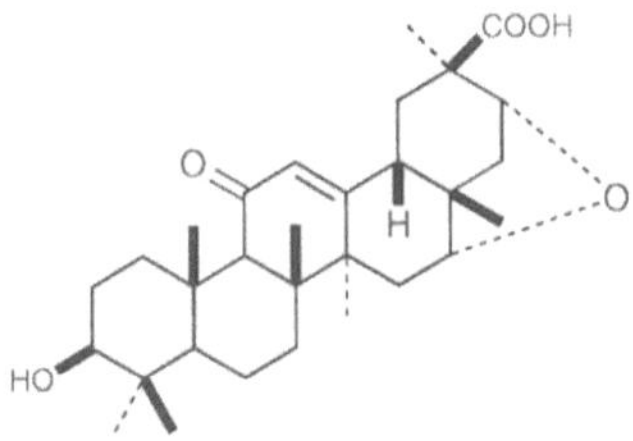

liquoric acid (**6**)

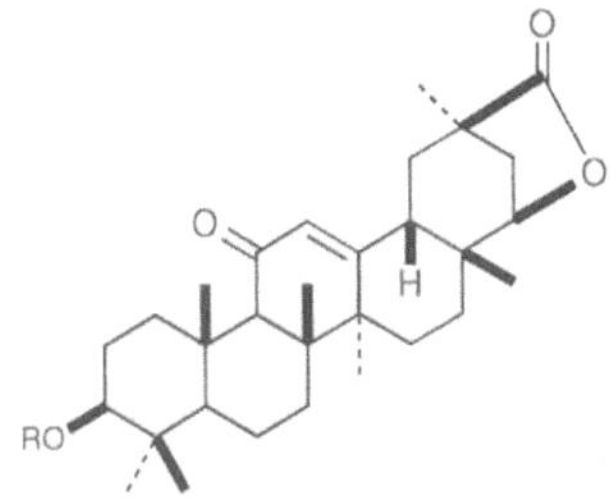

glabrolide (**7**) : R = H
(**27**) : R = -COH

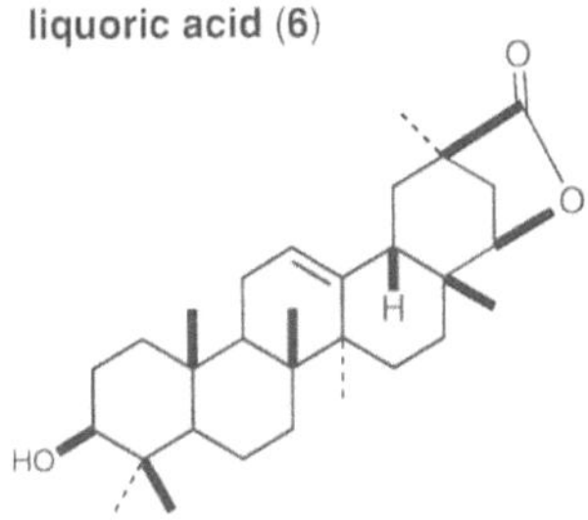

11-deoxoglabrolide (**8**)

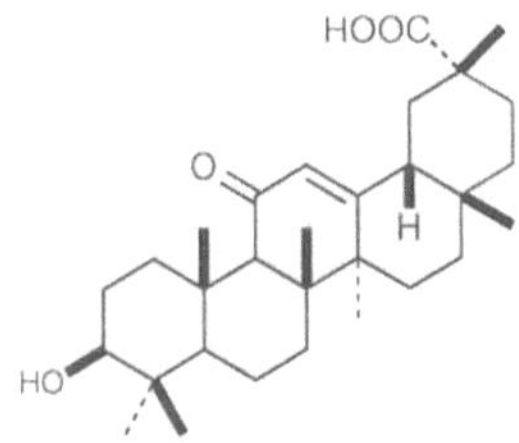

liquiritic acid (**9**)

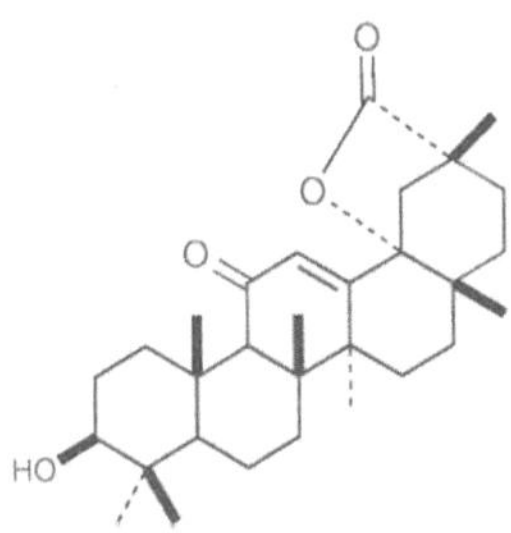

isoglabrolide (**10**)

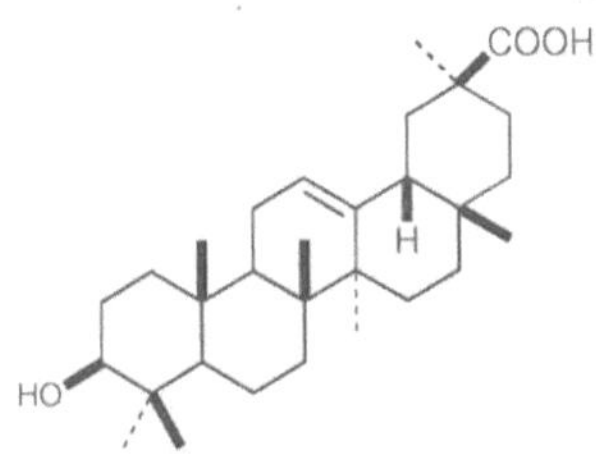

11-deoxoglycyrrhetinic acid (**11**)

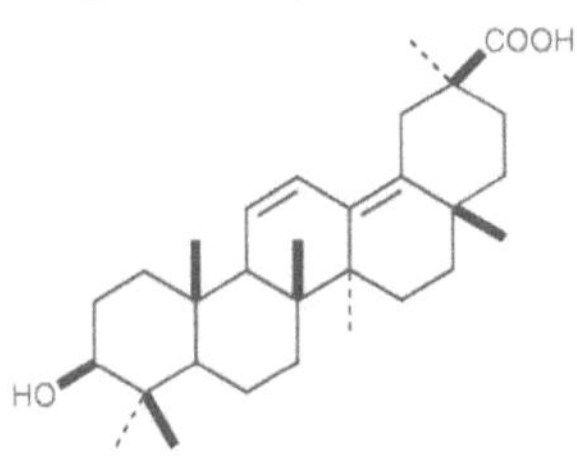

3β-hydroxy-11,13(18)-oleanadien-30-oic acid (**12**)
(**glypallidifloric acid**)

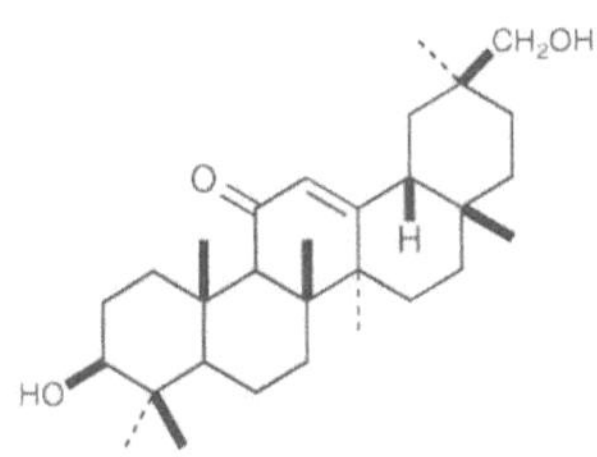

glycyrretol (**13**)
(**glycyrrhetol**)

Fig. 2

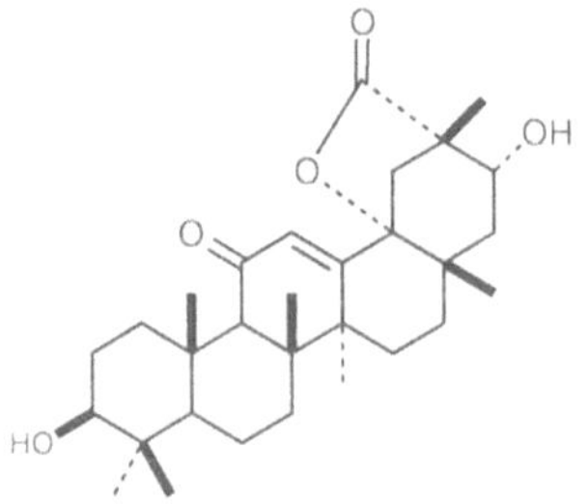

21α-hydroxyisoglabrolide (14)

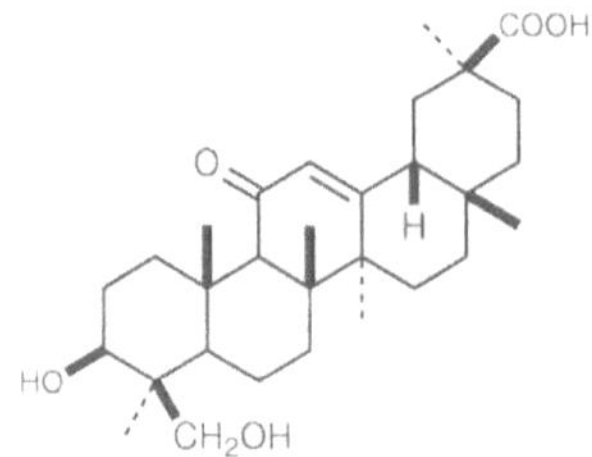

24-hydroxyglycyrrhetic acid (15)

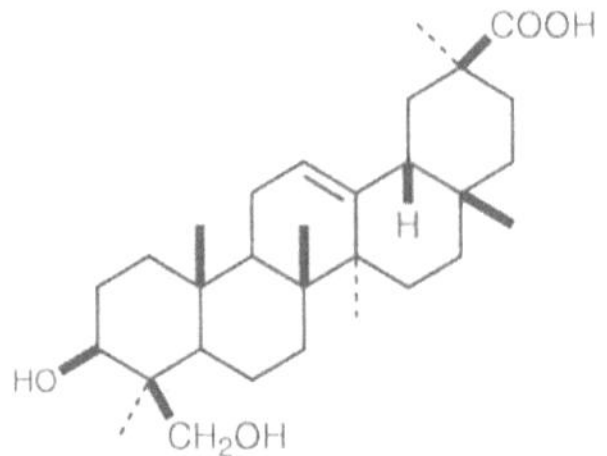

24-hydroxy-11-deoxo-
glycyrrhetic acid (16)

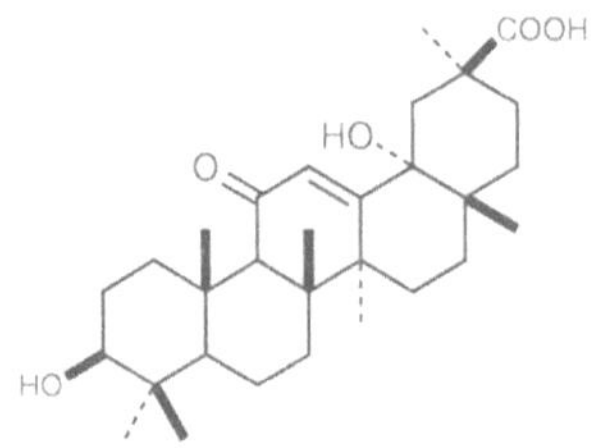

18α-hydroxyglycyrrhetic acid (17)

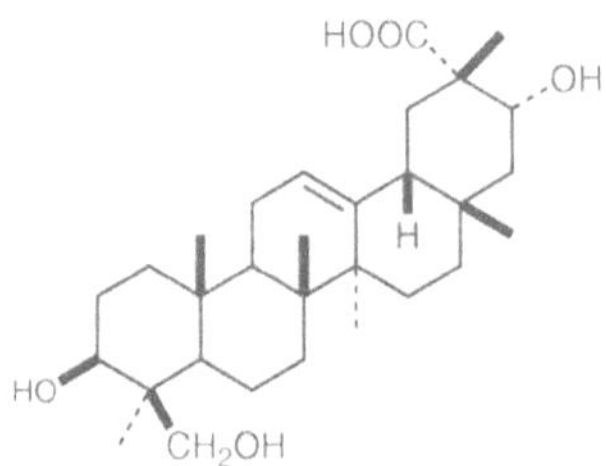

liquiridiolic acid (18)
(glyyunnansapogenin B_1)

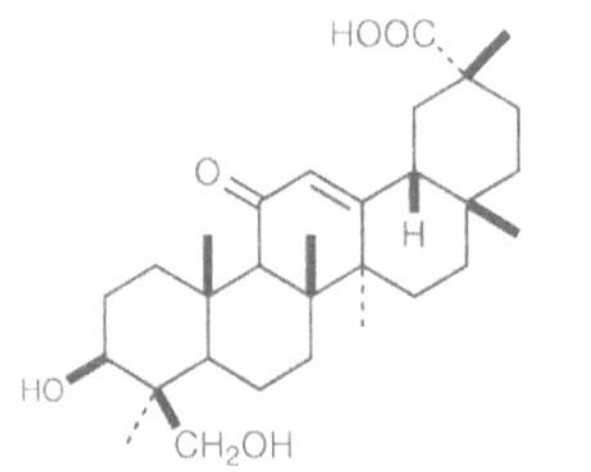

24-hydroxyliquiritic acid (19)

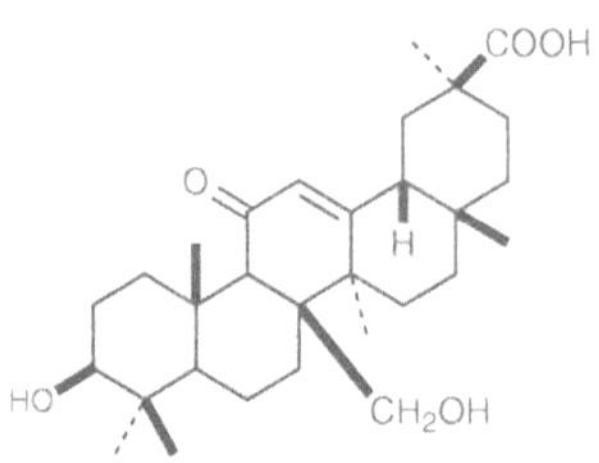

glabric acid (20)

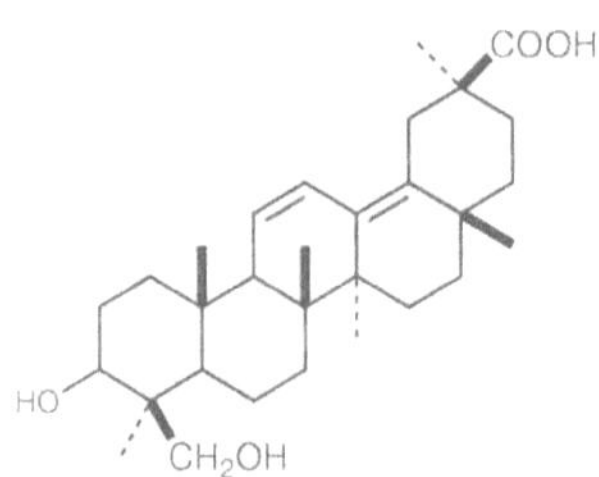

3,24-dihydroxy-11,13(18)-
oleanadien-30-oic acid (21)

Fig. 3

References, pp. 110–140

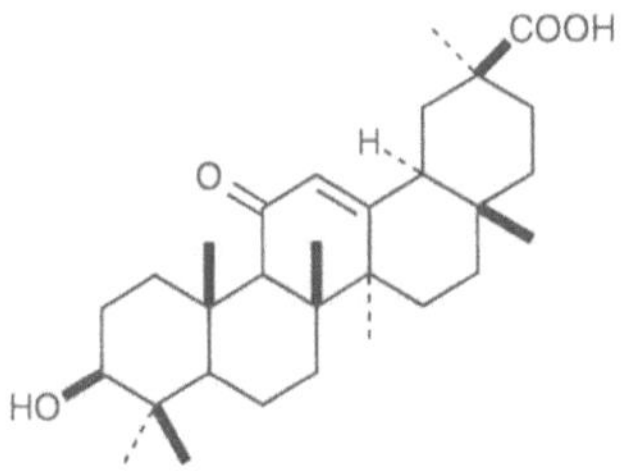

uralenic acid (22)

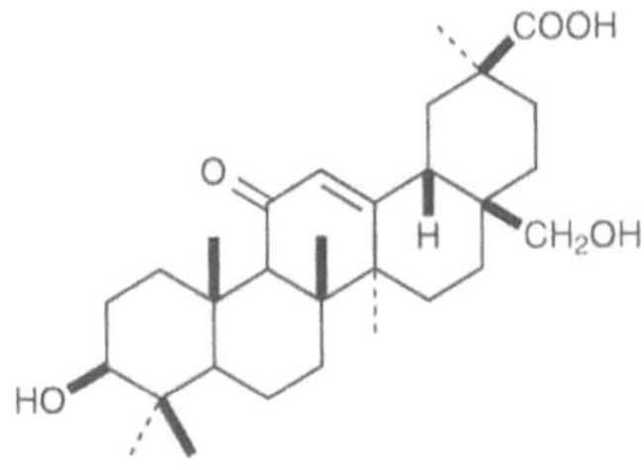

28-hydroxyglycyrrhetic acid (23)

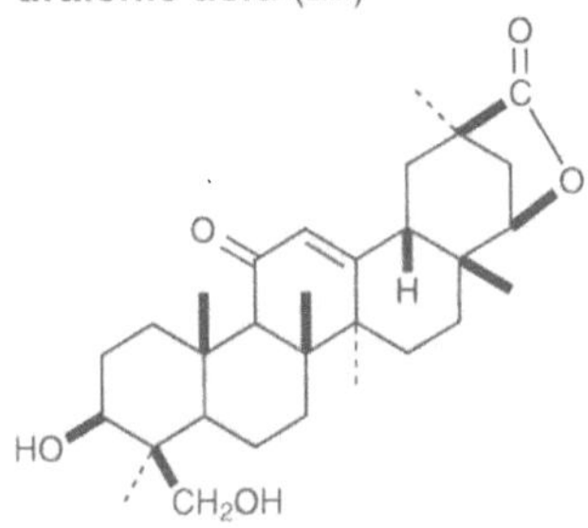

24-hydroxyglabrolide (24)

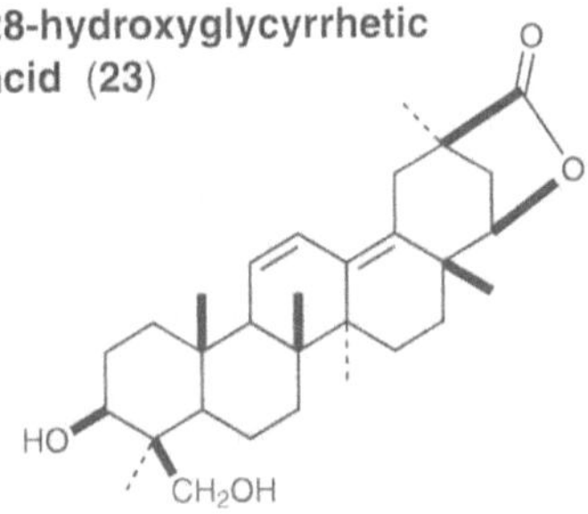

uralenolide (25)

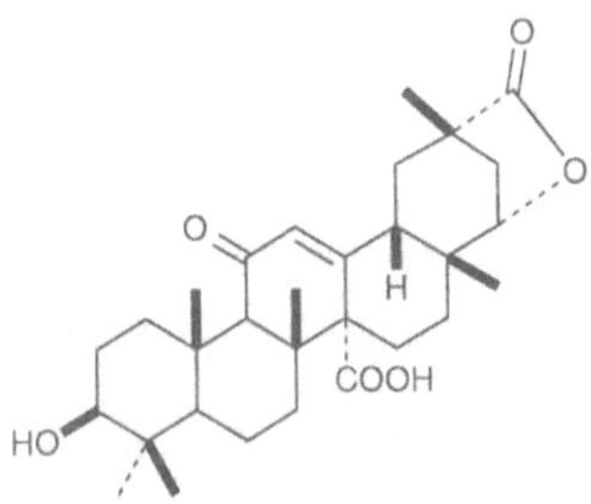

glyuranolide (26)

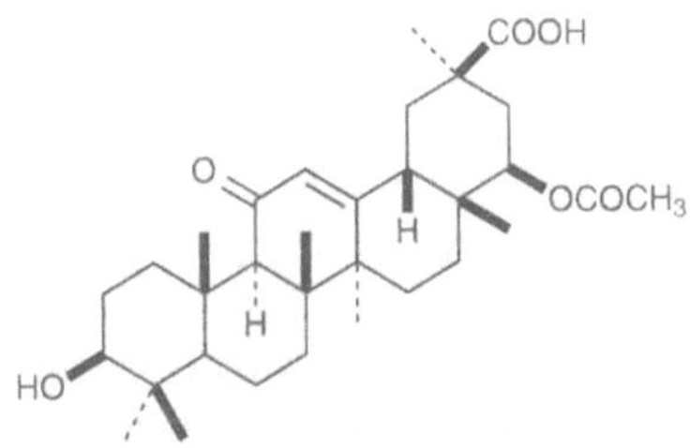

22β-acetoxyglycyrrhetic acid (28)

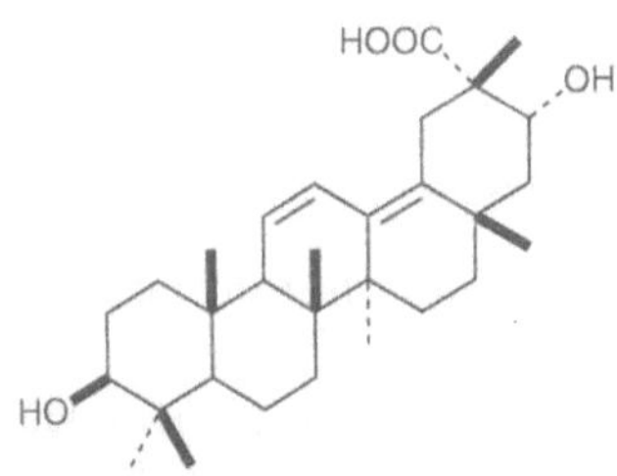

macedonic acid (29)

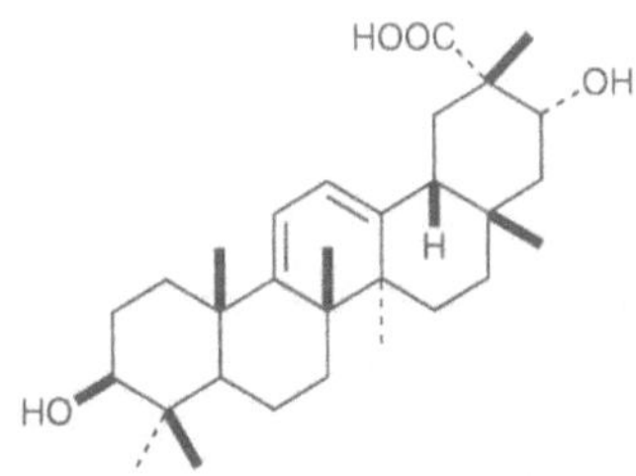

isomacedonic acid (30)

Fig. 4

2.2.3. *Glycyrrhiza echinata (G. macedonica)*

This species does not produce glycyrrhizic acid (*180*). The isolation of two triterpene acids named macedonic acid and isomacedonic acid whose structures were assigned as 3,19-dihydroxy-11,13(18)-oleanadien-28-oic acid and 3,19-dihydroxy-9(11),12-oleanadien-28-oic acid, respectively, was reported by KIR'YALOV *et al.* (*181–184*). However, these structures have been revised to **29** and **30**, respectively, using chemical and spectroscopic methods, by ZORINA *et al.* (*185, 186*). Similarly structures of two other acids, echinatic acid and isoechinatic acid, also isolated by KIR'YALOV *et al.* and assigned as 3,15,22-trihydroxy-9(11), 12-oleanadien-28-oic acid and 3,15,22 (or 3,16,21)-trihydroxy-11,13 (18)-oleanadien-28-oic acid, respectively, (*182, 187–189*) were subsequently revised to **31** and **32** (glyyunnansapogenin E), respectively (*190*). More recently the isolation of 3β,21α-dihydroxy-11,13(18)-oleanadien-28-oic acid (**33**) from the same species has been reported by MIRHOM *et al.* (*191*).

2.2.4. *Glycyrrhiza yunnanensis*

Glycyrrhetic acid has not been detected in this species (*36*), but six sapogenins, glyyunnansapogenins A–C (**34–36**) and F–H (**37–39**), were isolated by ZENG *et al.* (*192–195*). The isolation of glyyunnansapogenin B_2 (**40**) and 24-hydroxyglabrolide (**24**) has been reported by HU *et al.* (*67*). The known sapogenins oleanolic acid (**41**), **18** (glyyunnansapogenin B_1), **29**, **32** (glyyunnansapogenin E) and yunganogenin C (**42**) have also been isolated from this species (*67, 193*).

2.2.5. *Other Species*

Pallidifloric acid (**43**, as a methyl ester) has been isolated from *G. pallidiflora* by LIU *et al.* (*196*) as have the known sapogenins, **29**, **12** (glypallidifloric acid) and soyasapogenol B (**44**) (*196–199*). A known sapogenin glabrolide (**7**) was isolated from *G. aspera* (*200*) and 24-hydroxyglycyrrhetic acid (**15**) was obtained from *G. korshinskyi* (*201*). A new sapogenin squasapogenol (**45**) has been isolated together with two known sapogenins (**29** and **44**) from *G. squamulosa* by LIANG and ZHANG (*202*).

Several sapogenins (**46–55**) have been obtained as hydrolysis products of saponins from *G. uralensis* and *G. yunnanensis*. These are described in the next section.

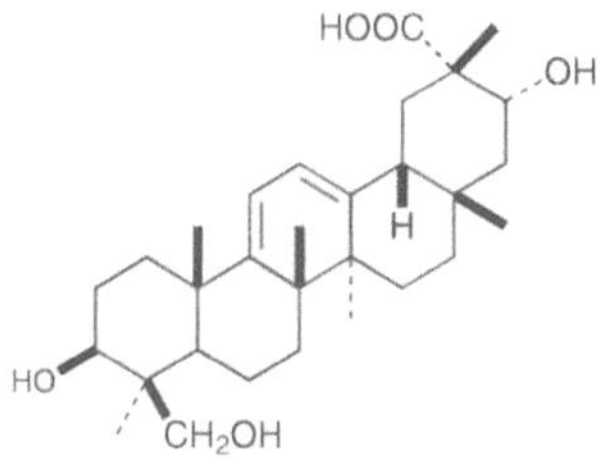

echinatic acid (31)

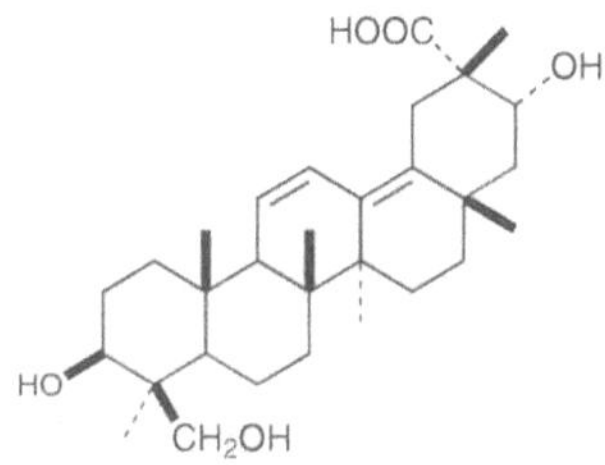

isoechinatic acid (32)
(glyyunnansapogenin E)

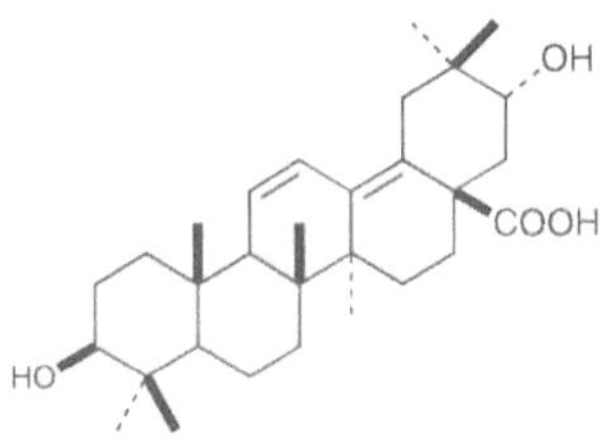

3β,21α-dihydroxy-11,13(18)-oleanadien-28-oic acid (33)

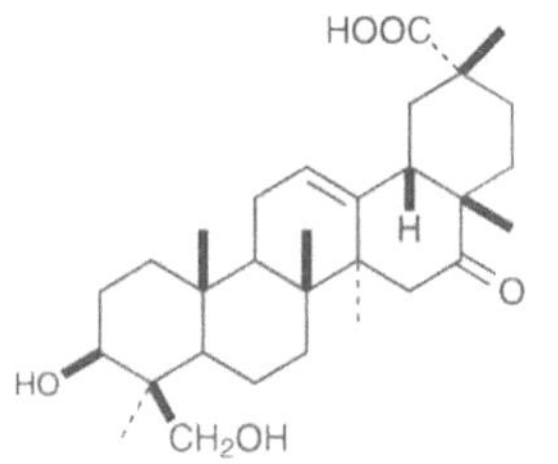

glyyunnansapogenin A (34)

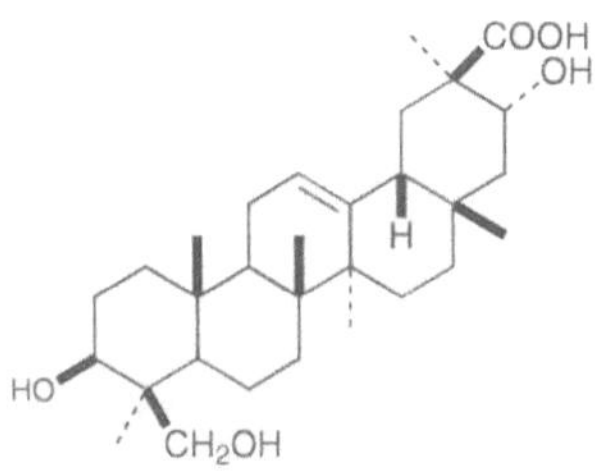

glyyunnansapogenin B (35)

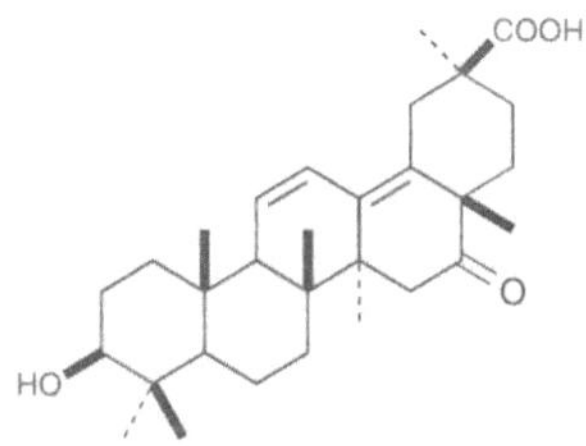

glyyunnansapogenin C (36)

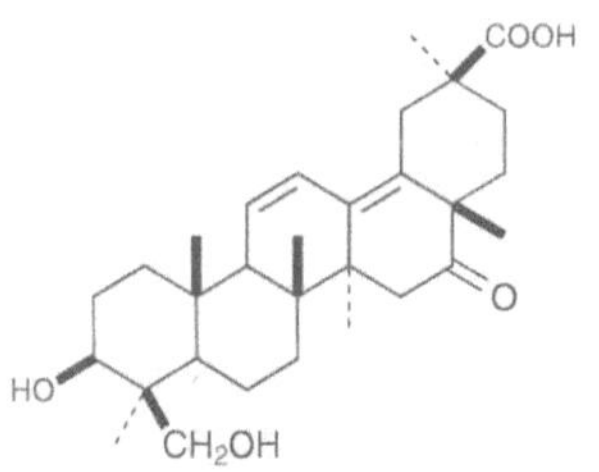

glyyunnansapogenin F (37)

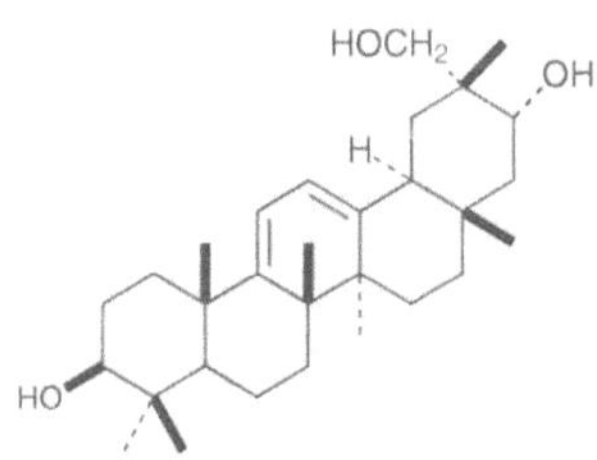

glyyunnansapogenin G (38)

Fig. 5

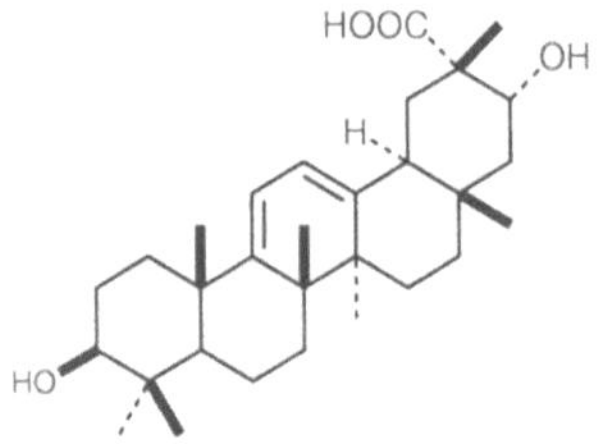

glyyunnansapogenin H (**39**)

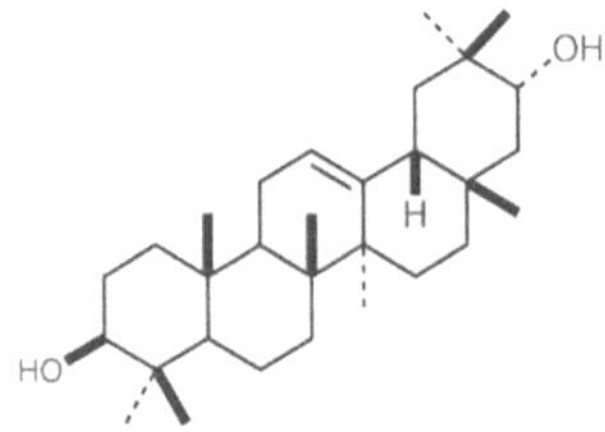

glyyunnansapogenin B_2 (**40**)

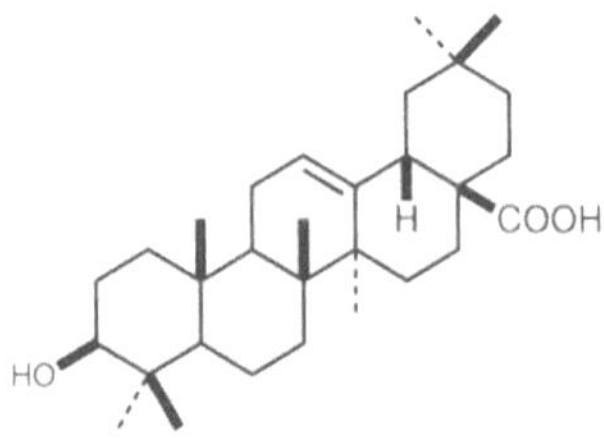

oleanolic acid (**41**)

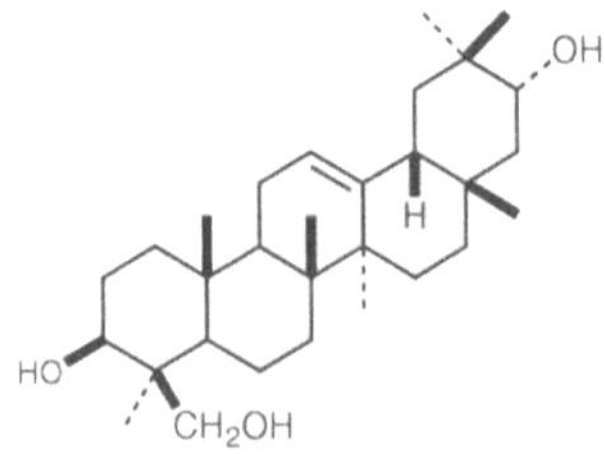

yunganogenin C (**42**)

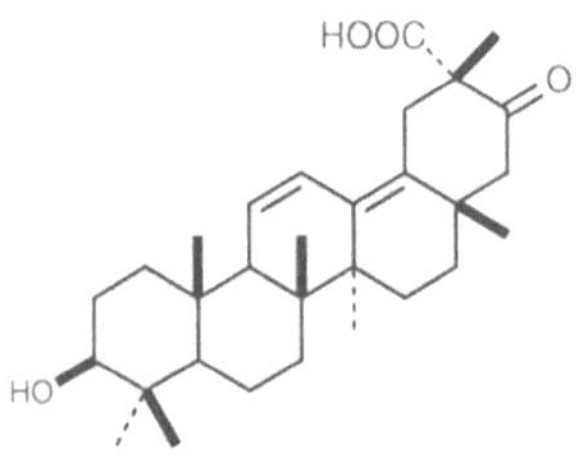

pallidifloric acid (**43**)

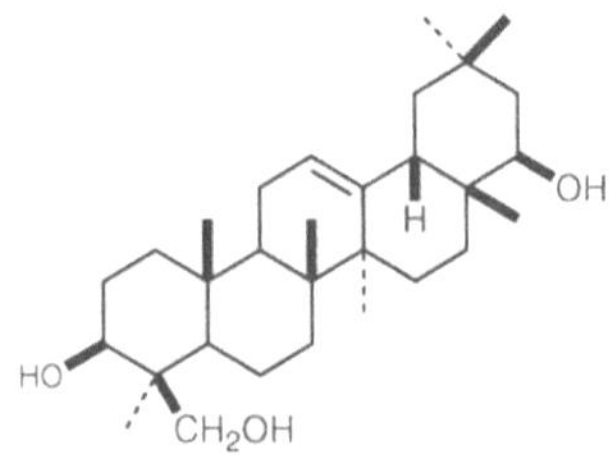

soyasapogenol B (**44**)

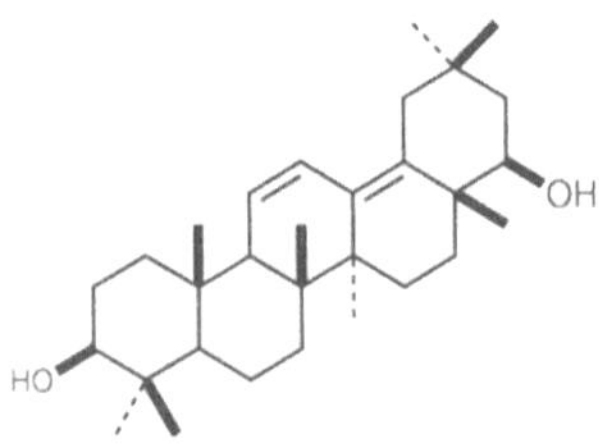

squasapogenol (**45**)

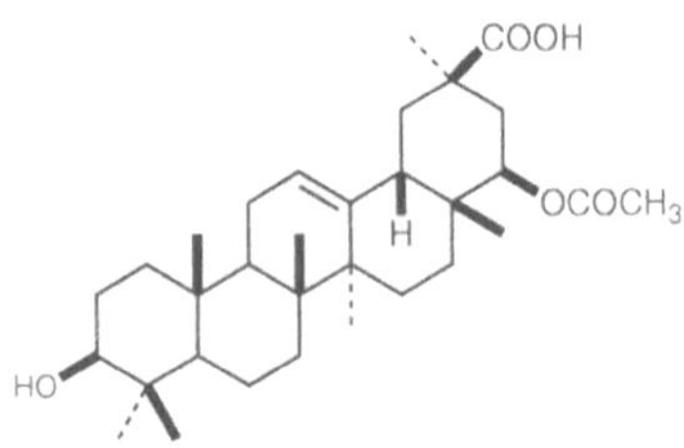

22β-acetoxy-3β-hydroxy-12-oleanene-30-oic acid (**46**)

Fig. 6

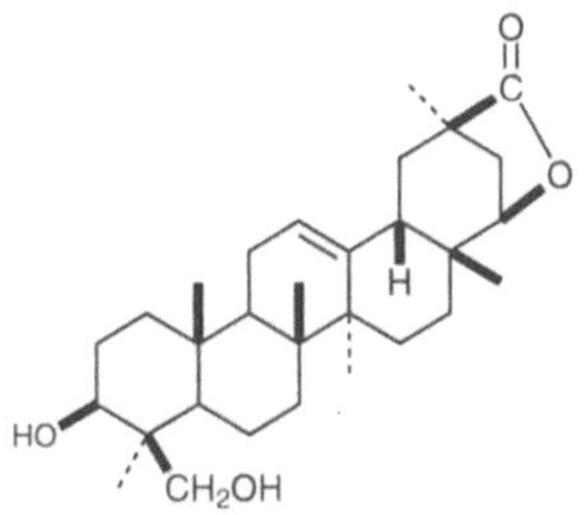

24-hydroxy-11-deoxo-glabrolide (47)

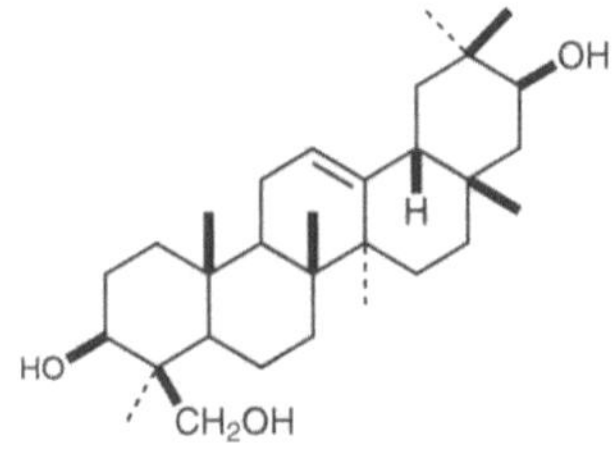

kudzusapogenol C (48)

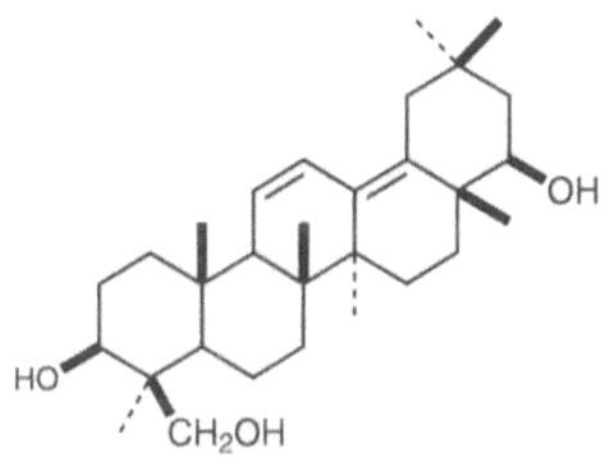

yunganogenin D (49)

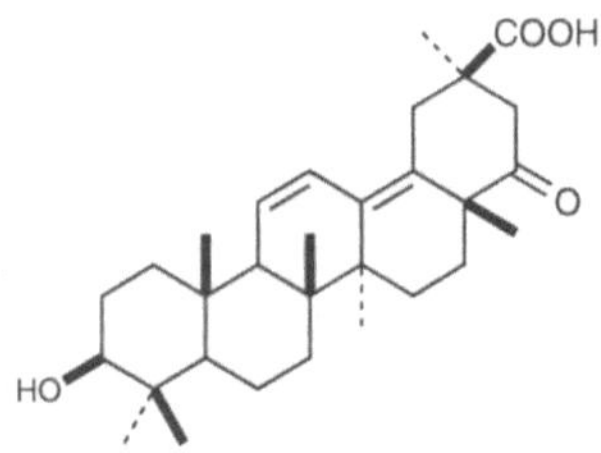

yunganogenin E (50)

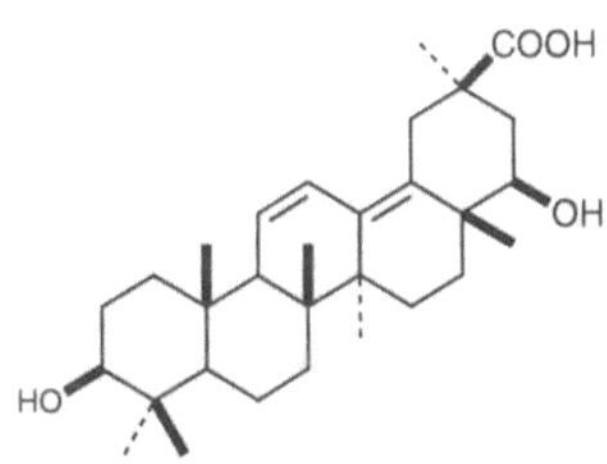

yunganogenin F (51)

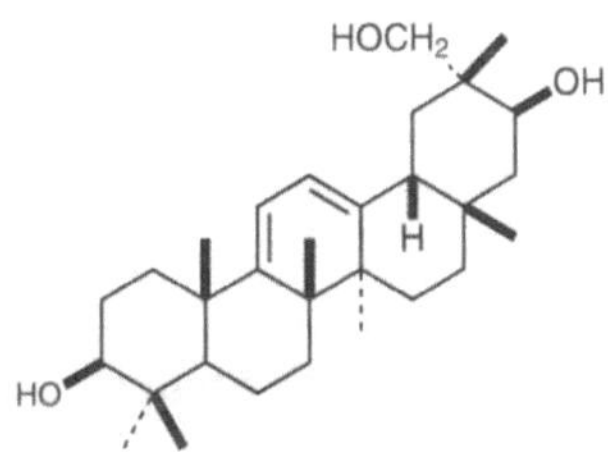

yunganogenin H (52)

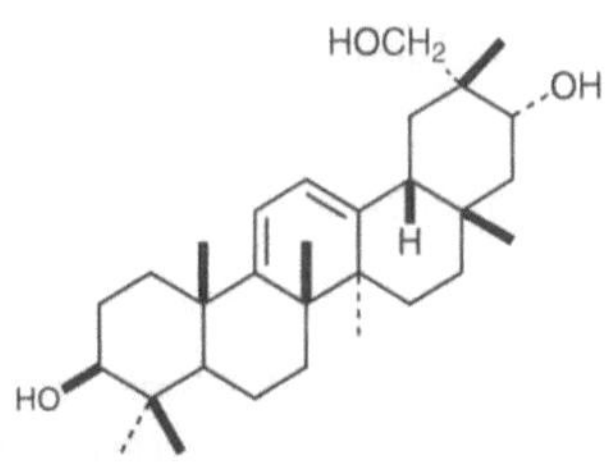

yunganogenin I (53)

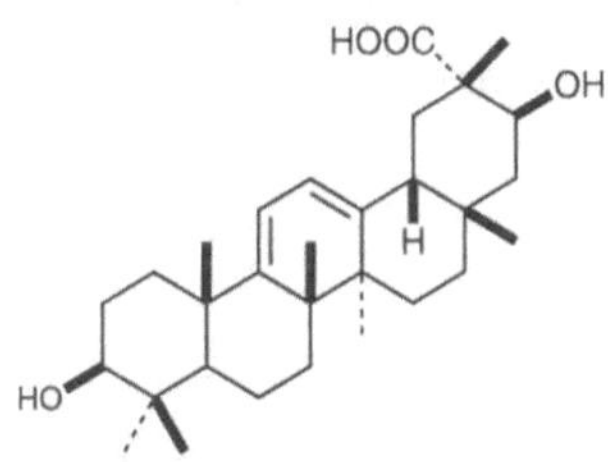

yunganogenin J (54)

Fig. 7

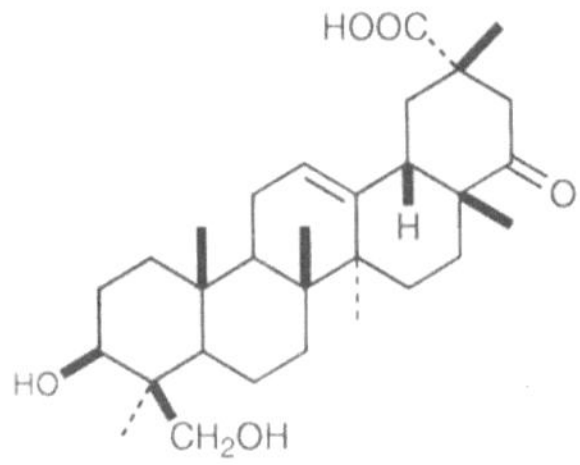

yunganogenin K (55)

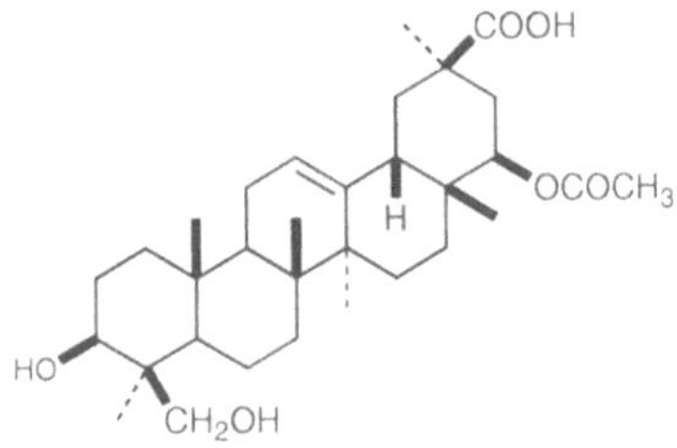

22β-acetoxy-3β,24-dihydroxy-12-oleanene-30-oic acid (56)*

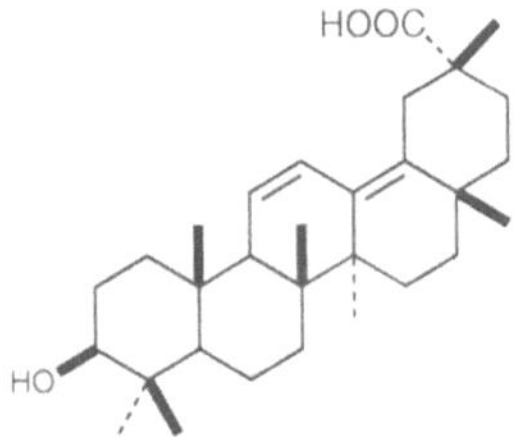

3β-hydroxy-11,13(18)-oleanadiene-9-oic acid (57)

Fig. 8. * The compound has not been isolated (aglycone moiety of **68** or **82**)

2.3. Structures of Minor Saponins

About fifty saponins have so far been isolated from *Glycyrrhiza* species. Their structures are shown in Table 2, and their physical properties and plant sources are described in Table 12.

2.3.1. Glycyrrhiza uralensis

KITAGAWA *et al.* have reported the isolation of a new bisdesmoside, licorice-saponin A3 (**58**), and ten new monodesmosides, licorice-saponins B2, C2, D3, E2, F3, G2, H2, J2, K2, L3 (**59–68**) together with the known saponins glycyrrhizic acid (**2**) and 18α-glycyrrhizin (**69**) from *G. uralensis* (*100, 203–206*). Methanolysis of the trimethyl ester of licorice-saponin D3 (**61a**) liberates 22β-acetoxy-3β-hydroxy-12-oleanene-30-oic acid (**46**) as a methyl ester, and subsequent hydrolysis gives 11-deoxoglabrolide (**8**) (*205*). A similar triterpenoid lactone (24-hydroxy-11-deoxoglabrolide, **47**) is also obtained by methanolysis of licorice-saponin L3 (**68**) (*206*). Among these saponins, licorice-saponin

B2 (**59**) has been noted to exhibit activity which promotes restoration of hepatocyte function against carbon tetrachloride induced liver cell injury in rat (*206*).

That licorice may act as a solubilizing agent in traditional Chinese medicines has been reported by SASAKI *et al.* (*207*). Aqueous solutions of the glucoside ester (**58**) solubilize water insoluble substances such as α-tocopherol and oleanolic acid. This may be a restatement of a description of licorice in the "Shen Nong Ben Cao Jing," which states that licorice harmonizes a medicine with any drugs.

Uralsaponin B (**70**) has been isolated from *G. uralensis* by ZHANG *et al.* (*127*). The saponin is a positional isomer of glycyrrhizic acid (**2**) with respect to the attachment of the outer glucuronyl group.

2.3.2. Glycyrrhiza eurycarpa

Another such isomer of glycyrrhizic acid (**2**) differing in the attachment of the glucuronyl group, glyeurysaponin (**71**) has been isolated from *G. eurycarpa* by ZHANG *et al.* (*208*).

2.3.3. Glycyrrhiza inflata

Apioglycyrrhizin (**72**) and araboglycyrrhizin (**73**) have been isolated from *G. inflata* by KITAGAWA *et al.* (*209, 210*) who have also examined the taste of licorice saponins obtained from various species: Apioglycyrrhizin (**72**) is about 2 times sweeter than glycyrrhizic acid (**2**), while the sweetness of araboglycyrrhizin (**73**) and the licorice-saponins G2 (**64**) or H2 (**65**) is comparable to that of **2**. On the other hand, licorice-saponin B2 (**59**) is not sweet but bitter and causes vomiting. The sweetness of licorice-saponin A3 (**58**) is weaker by about half than that of **2**, but imparts a more pleasant taste. However SASAKI *et al.* reported that saponin **58** derived from **2** and glycyrrhizic acid 30-*O*-β-glucronide ester are tasteless (*207*). The isolation of methyl (ethyl) and/or *n*-butyl diglucuronides of glycyrrhetic acid or liquiritic acid, inflasaponins I–VI (**74**–**78**), from *G. inflata* has been reported by ZHANG *et al.* (*211–214*).

2.3.4. Glycyrrhiza glabra

VARSHENY *et al.* have reported the isolation of a pentaglycoside of glycyrrhetic acid, glabranin A (**79**),* and a heptaglycoside, glabranin B

* The trivial name "glabranin" is also given to a flavanone **267** as described in Section 3 (*226*).

(**80**), from *G. glabra*, but the structures were not established completely (*215*).

2.3.5. *Glycyrrhiza yunnanensis*

The isolation of a bisdesmoside, glyyunnanprosapogenin D (**81**) as a dimethyl ester hexaacetate, and two monodesmosides, yunnanglysaponins A (**82**) and B (**83**), from *G. yunnanensis* has been reported by ZHANG *et al.* (*194, 216*). Eighteen new saponins, yunganosides A1, B1, C1, D1, E2, F2, G1, G2, H1, H2, I1, I2, J1, J2, K1, K2, L1 and L2 (**84**–**101**) have been isolated by TANAKA *et al.*, and eight new aglycones, yunganogenins C–F (**42, 49**–**51**, Fig. 7) and H–K (**52–55**, Figs. 7 and 8), obtained by the hydrolysis of the corresponding saponins with glycyrrhizin hydrolase were also reported by them (*217, 218*).*

2.3.6. *Other Species*

Macedonic acid diglucuronide (**102**) has been isolated from *G. echinata* and *G. pallidiflora* by KUSANO *et al.* (*219*). The saponin also occurs in *G. macedonica*, but not in *G. uralensis* and *G. glabra* (private communication from Dr. KUSANO; he and his co-workers consider that *G. macedonica* is a separate species from *G. echinata*).

2.3.7. *Biotransformations with Plant Tissue Cultures*

The biotransformation of glycyrrhetic acid (**1**) with cell cultures of *G. glabra* has been reported by TABATA *et al.* (*220, 221*). The cell suspension cultures convert exogenously administered glycyrrhetic acid (**1**) into monodesmosides (**103**–**107**) and a bisdesmoside (**108**). These saponins have not been isolated from *Glycyrrhiza* species. Some of the metabolites (**103**–**105**) are hydroxylated at the C-24 methyl group of the aglycone as well as of the sapogenin, 24-hydroxyglycyrrhetic acid (**15**), also obtained from the same cell line. In the cell line, 24-hydroxylation of glycyrrhetic acid is pre-requisite to forming its 3-*O*-glucuronide (*222*).

The biotransformation of papaverine with cell suspension cultures of *G. glabra* has been reported by DORISSE *et al.* (*223*). The main metabolite of the transformation is a hydroxylated compound, papaverinol. The

* The authors also stated that the glycosides of sapogenins **32** or **36** reported by ZENG *et al.* from *G. yunnanensis* (*193*), were not encountered by them (*217*).

Table 2. *Structure of Saponins (GlcA = Glucuronopyranosyl, Glu = Glucopyranosyl, Api = Apiofuranosyl, Ara = Arabinopyranosyel, Gal = Galactopyranosyl, Rha = Rhamnopyranosyl, Xyl = Xylopyranosyl)*

Glycyrrhizic acid (**2**):
β-D-GlcA 1 → 2 β-D-GlcA 1 → 3 [glycyrrhetic acid (**1**)]

Soyasaponin I (**3**):
α-L-Rha 1 → 2 β-D-Gal 1 → 2 β-D-GlcA 1 → 3 [soyasapogenol B (**44**)]

Soyasaponin II (**4**):
α-L-Rha 1 → 2 α-L-Ara 1 → 2 β-D-GlcA 1 → 3 [soyasapogenol B (**44**)]

Licorice-saponin A3 (**58**):
β-D-GlcA 1 → 2 β-D-GlcA 1 → 3 [glycyrrhetic acid (**1**)] 30 ← 1 β-D-Glu

Licorice-saponin B2 (**59**):
β-D-GlcA 1 → 2 β-D-GlcA 1 → 3 [11-deoxoglycrrhetinic acid (**11**)]

Licorice-saponin C2 (**60**):
β-D-GlcA 1 → 2 β-D-GlcA 1 → 3 [3β-hydroxy-11,13(18)-oleanadiene-30-oic acid (**12**)]

Licorice-saponin D3 (**61**):
α-L-Rha 1 → 2 β-D-GlcA 1 → 2 β-D-GlcA 1 → 3 [22β-acetoxy-3β-hydroxy-12-oleanene-30-oic acid (**46**)]

Licorice-saponin E2 (**62**):
β-D-GlcA 1 → 2 β-D-GlcA 1 → 3 [glabrolide (**7**)]

Licorice-saponin F3 (**63**):
α-L-Rha 1 → 2 β-D-GlcA 1 → 2 β-D-GlcA 1 → 3 [11-deoxoglabrolide (**8**)]

Licorice-saponin G2 (**64**):
β-D-GlcA 1 → 2 β-D-GlcA 1 → 3 [24-hydroxyglycyrrhetic acid (**15**)]

Licorice-saponin H2 (**65**):
β-D-GlcA 1 → 2 β-D-GlcA 1 → 3 [liquiritic acid (**9**)]

Licorice-saponin J2 (**66**):
β-D-GlcA 1 → 2 β-D-GlcA 1 → 3 [24-hydroxy-11-deoxoglycyrrhetinic acid (**16**)]

Licorice-saponin K2 (**67**):
β-D-GlcA 1 → 2 β-D-GlcA 1 → 3 [3β,24-dihydroxy-11,13(18)-oleanadiene-30-oic acid]

Licorice-saponin L3 (**68**):
α-L-Rha 1 → 2 α-L-Ara 1 → 2 β-D-GlcA 1 → 3[22β-acetoxy-3β,24-dihydroxy-12-oleanene-30-oic acid (**56**)[a]]

18α-Glycyrrhizin (**69**):
β-D-GlcA 1 → 2 β-D-GlcA 1 → 3 [uralenic acid (**22**)]

Uralsaponin B (**70**):
β-D-GlcA 1 → 3 β-D-GlcA 1 → 3 [glycyrrhetic acid (**1**)]

Table 2 (*continued*)

Glyeurysaponin (**71**):

β-D-GlcA 1 → 4 β-D-GlcA 1 → 3 [glycyrrhetic acid (**1**)]

Apioglycyrrhizin (**72**):

β-D-Api 1 → 2 β-D-GlcA 1 → 3 [glycyrrhetic acid (**1**)]

Araboglycyrrhizin (**73**):

α-L-Ara 1 → 2 β-D-GlcA 1 → 3 [glycyrrhetic acid (**1**)]

Inflasaponin I (**74**):

CH_3- → 6 β-D-GlcA 1 → 2 β-D-GlcA 1 → 3 [glycyrrhetic acid (**1**)]
C_4H_9- ↑ 6

Inflasaponin II (**75a**):

C_2H_5- → 6 β-D-GlcA 1 → 2 β-D-GlcA 1 → 3 [glycyrrhetic acid (**1**)]
n-C_4H_9- ↑ 6

Inflasaponin III (**76**):

n-C_4H_9- → 6 β-D-GlcA 1 → 2 β-D-GlcA 1 → 3 [liquiritic acid (**9**)]
C_2H_5- ↑ 6

Inflasaponin IV (**77**):

n-C_4H_9- → 6 β-D-GlcA 1 → 2 β-D-GlcA 1 → 3 [glycyrrhetic acid (**1**)]
n-C_4H_9- ↑ 6

Inflasaponin V (**78**):

n-C_4H_9- → 6 β-D-GlcA 1 → 2 β-D-GlcA 1 → 3 [liquiritic acid (**9**)]
n-C_4H_9- ↑ 6

Inflasaponin VI (**75b**):

n-C_4H_9- → 6 β-D-GlcA 1 → 2 β-D-GlcA 1 → 3 [glycyrrhetic acid (**1**)]
C_2H_5- ↑ 6

Glabranin A (**79**): (partial structure)

α-L-Rha → α-D-Glu × 2 → α-D-Glu 1 → 3 [glycyrrhetic acid (**1**)] + α-D-Glu

Glabranin B (**80**): (partial structure)

(α-D-Glu ×2) 1 → α-D-Glu 1 → 3 [glycyrrhetic acid (**1**)]
+ β-D-Xyl (terminal) → α-L-Rha × 2 →, and α-D-Glu (terminal) →
(hydrolyzed products of a methyl ether: 2,3,4-tri-*O*-methyl-β-D-xylose, 3,4-di-*O*-methyl-α-l-rhamnose, 2,3,4,5-tetra-*O*-methyl-α-D-glucose, 2,3,4-tri-*O*-methyl-α-D-glucose, 2,3-di-*O*-methyl-α-D-glucose and **1**)

Glyyunnanprosapogenin D (**81**):

β-D-GlcA 1 → 3 [macedonic acid (**29**)] 21 ← 1 β-D-GlcA

Yunnanglysaponin A (**82**):

β-D-GlcA 1 → 4 β-D-GlcA 1→ 3 [3β-hydroxy-11,13(18)-oleanadien-29-oic acid (**57**)[a]]

Yunnanglysaponin B (**83**):

β-D-GlcA 1 → 4 β-D-GlcA 1 → 3 [macedonic acid (**29**)]

Yunganoside A1 (**84**):

α-L-Rha 1 → 2 β-D-GlcA 1 → 2 β-D-GlcA 3 [kudzusapogenol C (**48**)]

Table 2 (*continued*)

Yunganoside B1 (**85**):

α-L-Rha 1 → 2 β-D-GlcA 1 → 2 β-D-GlcA 3 [soyasapogenol B (**44**)]

Yunganoside C1 (**86**):

α-L-Rha 1 → 2 β-D-GlcA 1 → 2 β-D-GlcA 3 [yunganogenin C (**42**)]

Yunganoside D1 (**87**):

α-L-Rha 1 → 2 β-D-GlcA 1 → 2 β-D-GlcA 3 [yunganogenin D (**49**)]

Yunganoside E2 (**88**):

β-D-GlcA 1 → 2 β-D-GlcA 1 → 2 β-D-GlcA 3 [yunganogenin E (**50**)]

Yunganoside F2 (**89**):

β-D-GlcA 1 → 2 β-D-GlcA 1 → 2 β-D-GlcA 3 [yunganogenin F (**51**)]

Yunganoside G1 (**90**):

α-L-Rha 1 → 2 β-D-GlcA 1 → 2 β-D-GlcA 1 → 3 [liquiridiolic acid (**18**)]

Yunganoside G2 (**91**):

β-D-GlcA 1 → 2 β-D-GlcA 1 → 3 [liquiridiolic acid (**18**)]

Yunganoside H1 (**92**):

α-L-Rha 1 → 2 β-D-GlcA 1 → 2 β-D-GlcA 1 → 3 [yunganogenin H (**52**)]

Yunganoside H2 (**93**):

β-D-GlcA 1 → 2 β-D-GlcA 1 → 3 [yunganogenin H (**52**)]

Yunganoside I1 (**94**):

α-L-Rha 1 → 2 β-D-GlcA 1 → 2 β-D-GlcA 1 → 3 [yunganogenin I (**53**)]

Yunganoside I2 (**95**):

β-D-GlcA 1 → 2 β-D-GlcA 1 → 3 [yunganogenin I (**53**)]

Yunganoside J1 (**96**):

α-L-Rha 1 → 2 β-D-GlcA 1 → 2 β-D-GlcA 1 → 3 [yunganogenin J (**54**)]

Yunganoside J2 (**97**):

β-D-GlcA 1 → 2 β-D-GlcA 1 → 3 [yunganogenin J (**54**)]

Yunganoside K1 (**98**):

α-L-Rha 1 → 2 β-D-GlcA 1 → 2 β-D-GlcA 1 → 3 [yunganogenin K (**55**)]

Yunganoside K2 (**99**):

β-D-GlcA 1 → 2 β-D-GlcA 1 → 3 [yunganogenin K (**55**)]

Yunganoside L1 (**100**):

α-L-Rha 1 → 2 β-D-GlcA 1 → 2 β-D-GlcA 1 → 3 [isomacedonic acid (**30**)]

Yunganoside L2 (**101**):

β-D-GlcA 1 → 2 β-D-GlcA 1 → 3 [isomacedonic acid (**30**)]

Macedonic acid diglucronide (**102**):

β-D-GlcA 1 → 2 β-D-GlcA 1 → 3 [macedonic acid (**29**)]

3-*O*-β-D-Glucuronopyranosyl-24-hydroxy-18β-glycyrrhetinic acid (**103**):[b]

β-D-GlcA 1 → 3 [24-hydroxyglycyrrhetic acid (**15**)]

Table 2 (*continued*)

3-*O*-[α-L-Arabinopyranosyl-(1 → 2)-β-D-glucuronopyranosyl]-24-hydroxy-18β-glycyrrhetinic acid (**104**):[b]
[α-L-Ara] 1 → 2 [β-D-GlcA] 1 → 3 [24-hydroxyglycyrrhetic acid (**15**)]
24-Hydroxy-18β-glycyrrhetinic acid 30-β-D-glucopyranosyl ester (**105**):[b]
[β-D-Glu] 1 → 30 [24-hydroxyglycyrrhetic acid (**15**)]
3-*O*-β-D-Glucopyranosyl-18β-glycyrrhetinic acid (**106**):[b]
[β-D-Glu] 1 → 3 [glycyrrhetic acid (**1**)]
18-β-Glycyrrhetinic acid 30-*O*-β-D-glucopyranosyl ester (**107**):[b]
[β-D-Glu] 1 → 30 [glycyrrhetic acid (**1**)]
3-*O*-β-D-Glucopyranosyl-18β-glycyrrhetinic acid-30-β-D-glucopyranosyl ester (**108**):[b]
[β-D-Glu] 1 → 3 [glycyrrhetic acid (**1**)] 30 ← 1 [β-D-Glu]

[a] The compounds have been isolated as saponin but not obtained as free form
[b] Products with biotransformation

biotransformation of glycyrrhetic acid (**1**) by ginseng hairy root cultures has been reported by FURUYA *et al.* (*224*). The products of the biotransformation are three new glycosides and three known sapogenins. Among these compounds, 3-*O*-[β-D-glucopyranosyl-(1 → 2)-β-D-glucopyranosly]18β-glycyrrhetic acid and 3-*O*-[6-*O*-malonyl-β-glucopyranosly-(1 → 2)-β-D-glucopyranosyl]18β-glycyrrhetic acid possess the same degree of sweetness as glycyrrhizic acid (**2**).

3. Phenolic Compounds

3.1. Flavonoids and Coumarins from Underground Parts

Glycyrrhizic acid (**2**) and its derivative (**5**) are anti-ulcer drugs but have serious side effects, such as edema, reduced serum aldosterone, elevated blood pressure, *etc.* (*6*, *21*, *90*). In 1967, TAKAGI and his co-workers reported that one of the flavonoid-rich fractions of licorice which also included about 15% of glycyrrhizic acid (FM_{100} fraction) is effective in prevention of digestive gastric ulcer by suppressing gastric secretion (*24*, *227*).* The fraction has been developed as an anti-ulcer

* One of the semi-purified fractions obtained from the methanol extract of *G. glabra* (FM) had an anti-ulcer effect in rats at 100 mg/kg (i.p.). The same fraction from *G. uralensis* also exhibited a similar effect (*74*) and has been supplied as Aspalon®.

drug. In order to track down the biological activity, many studies of phenolic constituents have been conducted which are described in the following sections. About three hundred phenolic compounds have so far been isolated from *Glycyrrhiza* species, one hundred and fifty of which are new and have been obtained from the underground parts. Their structures are shown in Figs 9–20 and Scheme 1. Many known compounds have also been isolated from the roots or stolons. The structures and plant sources are described in Table 13 along with the new compounds.

3.1.1. Glycyrrhiza glabra

About seventy phenolic compounds have been isolated from the underground parts of this species. The main phenolic compound is isoliquiritin (**109**) in quantities of 0.1–0.3% from the dried roots (*228*). SHINODA and UEEDA have reported a flavanone glucoside liquiritin (**110**) from *G. glabra* (*229*, *230*). The corresponding chalcone, isoliquiritin (**109**), has been isolated from the dried roots of the plant together with flavanone glucoside (**110**) by PURI and SESHADRI (*231*). They also examined the fresh roots of the plant and obtained chalcone (**109**) but not flavanone (**110**). Apparently some isomerization of (**109**) to liquiritin (**110**) takes place because the flavanone glucoside was only detected on boiling the alcoholic extract or on exposure of the roots to sun drying or storage. On the other hand, the aglycone of liquiritin, liquiritigenin (**111**), has been isolated as an optical active compound from *G. glabra* as well as liquiritin (*232*, *233*). The corresponding chalcone, isoliquiritigenin (**112**), is also obtained from the plant (*232a*, *234*, *235*). Other glycosides based on the aglycones neoliquiritin (**113**), neoisoliquiritin (**114**), liquiritin apioside (apioliquiritin) (**115**), liquiritigenin 7-apiosylglucoside (**116**), rhamno-liquiritin (**117**), rhamno-isoliquiritin (**118**), licuraside (**119**) and licurazid (**120**) have been isolated from *G. glabra* (*232a*, *236–243*). Trivial names of the last two glycosides are not uniform or sometimes confused; isoliquiritigenin 4-*O*-apiosylglucoside (**119**, licuraside) has also been named neolicuroside or isoliquiritin apioside, while its isomer isoliquiritigenin 4′-*O*-apiosylglucoside (**120**, licurazid) has been called liquiraside, licuroside, licrasid, licuraside or licurasid (*210*, *232a*, *237*, *240–243*). Some of the multiplicity of names may have arisen as a result of typographical errors or as a result of translation into English; however, it may also be the result of the use of two different systems in numbering the chalcone skeleton (B ring, C1–C6 or C1′–C6′).

The oldest licorice in Japan derived from *G. glabra* has been stored in the Shosoin for over twelve hundred years (*244*). Shosoin is an Imperial

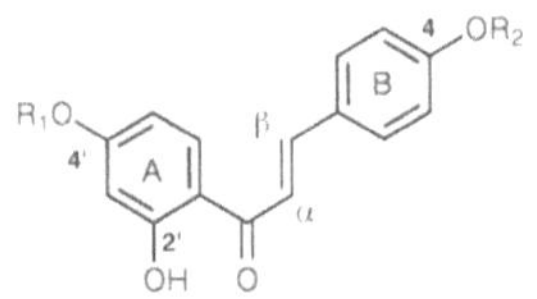

isoliquiritin (**109**):
R_1 = H, R_2 = Glu
isoliquiritigenin (**112**):
$R_1 = R_2$ = H
neoisoliquiritin (**114**):
R_1 = Glu, R_2 = H
rhamno-isoliquiritin (**118**):
R_1 = H, R_2 = Rha-Glu-,
licuraside (**119**):
R_1 = H, R_2 = Api(1 → 2)Glu-
licurazid (**120**):
R_1 = Api(1 → 2)Glu-, R_2 = H
(**159**): R_1 = H,
R_2 = (feruloyl →5)Api(1 →2)Glu-

feruloyl: 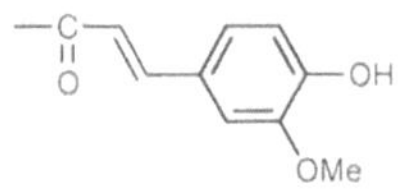

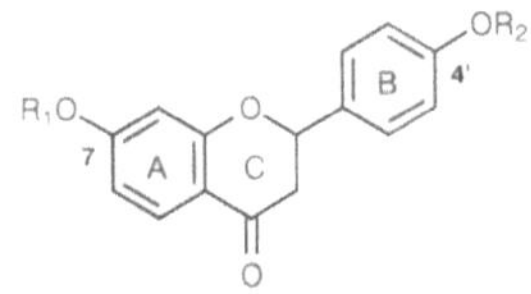

liquiritin (**110**):
R_1 = H, R_2 = Glu
liquiritigenin (**111**):
$R_1 = R_2$ = H
neoliquiritin (**113**):
R_1 = Glu, R_2 = H
liquiritin apioside (**115**):
R_1 = H, R_2 = Api(1 → 2)Glu-
liquiritigenin 7-apiosylglucoside (**116**):
R_1 = Api(1 → 2)Glu-, R_2 = H
rhamno-liquiritin (**117**):
R_1 = H, R_2 = Rha-Glu-
glucoliquiritin apioside (**149**):
R_1 = Glu, R_2 = Api(1 → 2)Glu-
liquiritigenin 4',7-diglucoside (**157**):
$R_1 = R_2$ = Glu
6"-acetylliquiritin (**158**):
R_1 = H, R_2 = (CH_3CO → 6)Glu

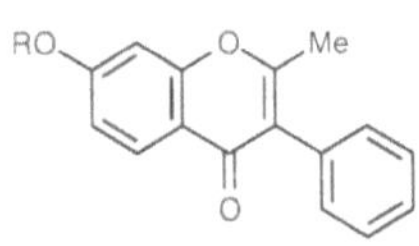

(**121**): R = H
(**121a**): R = CH_3CO

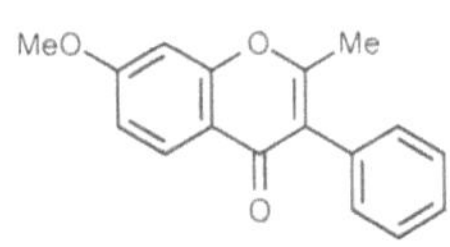

(**122**)

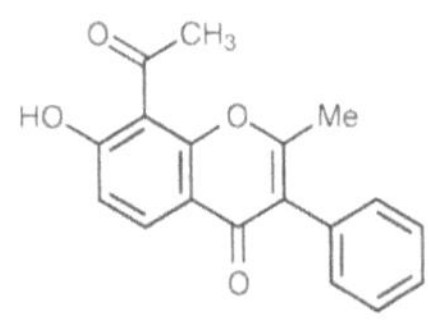

glyzarin (**123**)

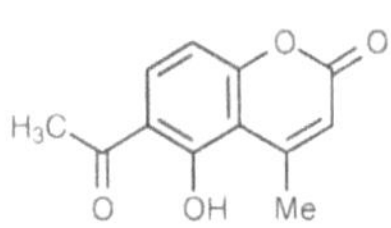

liqcoumarin (**124**)

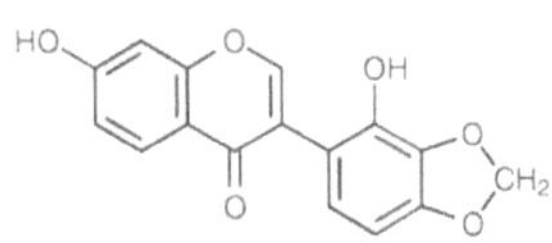

glyzaglabrin (**125**)

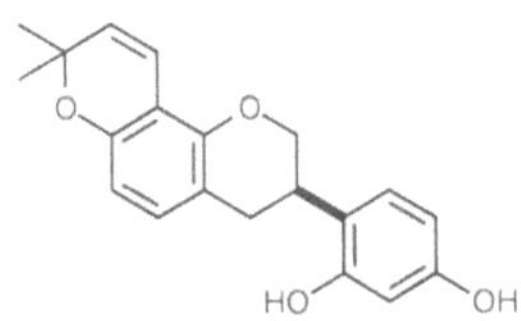

glabridin (**126**)

Fig. 9

Treasure House in the city of Nara, one of the ancient capital cities of Japan. The wooden house built in 756 A.D. contains many valuable items including sixty kinds of crude drugs. SHIBATA examined the constituents of this material using HPLC and detected the flavanones (**110**, **111**, **113**) and the chalcones (**109**, **112**, **114**) as well as glycyrrhetic acid (**1**) and glycyrrhizic acid (**2**) (*245*).*

Four 2-methylisoflavones (**121–123**, **121a**), and a coumarin, liqcoumarin (**124**), were isolated along with an isoflavone, glyzaglabrin (**125**), from licorice cultivated in India by Bhardwaj *et al.* (*246–249*). Isolation of these compounds from other licorice collections has not been reported.

Saitoh *et al.* have reported a pyranoisoflavan,† glabridin (**126**), and a flavanone with two prenyl groups, glabrol (**127**), isolated from Russian licorice (*250*). Two isoflavonoids having a 2,2-dimethylpyran ring attached to the B ring, an isoflavone (glabrone, **128**) and a pyranoisoflav-3-ene (glabrene), have been also obtained from the same source by Kinoshita *et al.* (*251*). Glabrene was originally assigned formula **129** on the basis of ^{1}H NMR data and a positive Gibbs test (*251*), but the structure was subsequently revised to formula **130** (*252*). Besides the flavonoids **126**, **127** and **130**, four new isoprenoid-substituted flavans, hispaglabridins A (**131**), B (**132**), 4′-*O*-methylglabridin (**133**) and "3′-methoxyglabridin (**134**)," have been isolated together with a 3-hydroxyflavanone, 3-hydroxyglabrol (**135**), from Spanish licorice in addition to salicylic acid and *O*-acetylsalicylic acid by Mitscher *et al.* (*253*, *254*). These flavonoids were also obtained from Russian licorice together with several new minor compounds, 8-prenyl-phaseolliniso-flavan (**136**) and compounds **137–142** (tentative names RL-P — -S, -U and -V) by Kinoshita *et al.* (*255–259*) who have also revised the structure of "3′-methoxyglabridin (**134**)" to 3′-hydroxy-4′-*O*-methyl-glabridin (**143**) (*255*). Among the isoprenoid-substituted flavonoids, glabridin (**126**) is the main constituent of Spanish and Russian licorice (about 0.15%), but the compound has not been obtained from *G. glabra* from northwestern China (*260–263*) except from material collected in the Inner Mongolia Autonomous Region where the content in the roots was 0.15% and in the rhizome 0.57% (*228*).

On examining the phenolic constituents of *G. glabra* cultivated in Japan we isolated two new pyrano-2-arylbenzofurans kanzonols U (**144**) and V (**145**), a pyrano-3-arylcoumarin kanzonol W (**146**), a diprenylated

---

* Dr. Shibata has informed us that he could not exclude the possibility of Shosoin licorice to be *G. uralensis* by the HPLC analysis and further chemical examination of the drug is now progress in 2nd research of Shosoin materials.

† The stereochemistry of isoflavan derivatives mentioned in this section is described in Section 4.

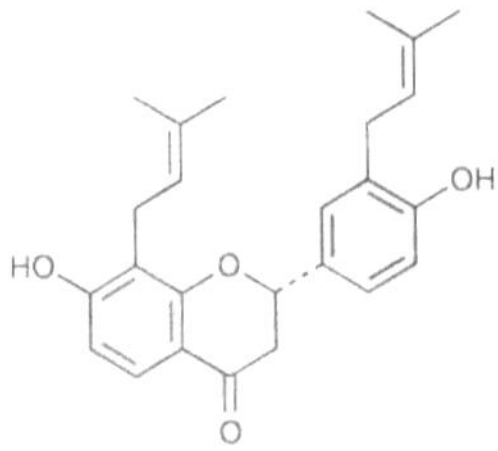

glabrol (127)

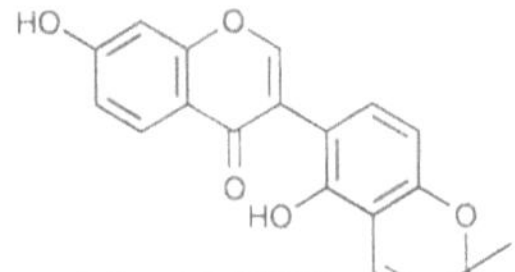

glabrone (128)

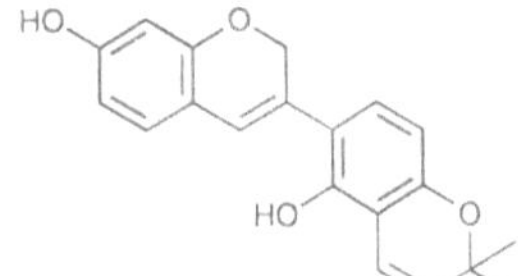

(129)

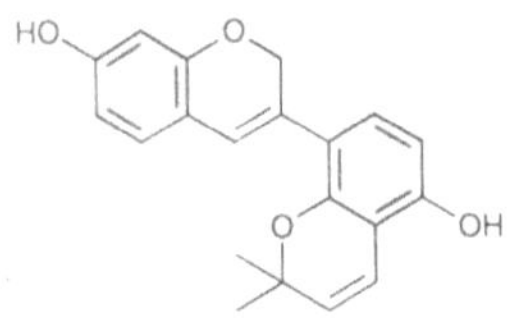

glabrene (130)

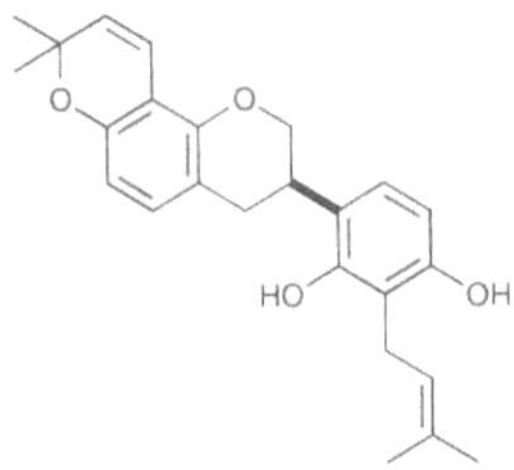

hispaglabridin A (131)

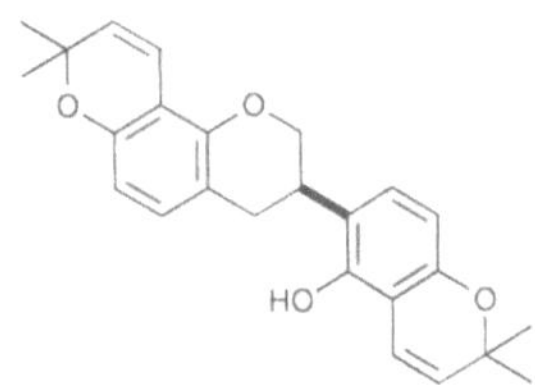

hispaglabridin B (132)

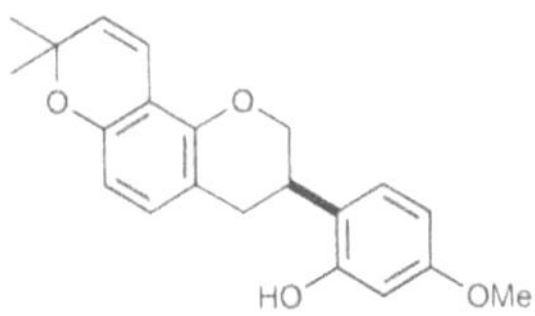

4'-*O*-methylglabridin (133)

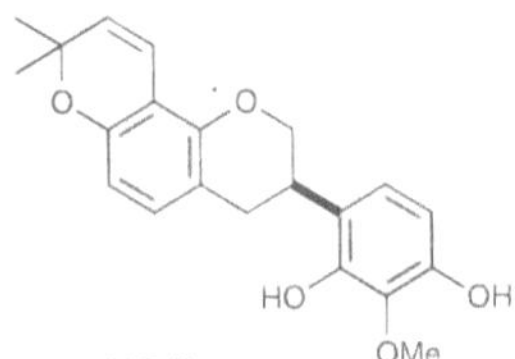

(134)

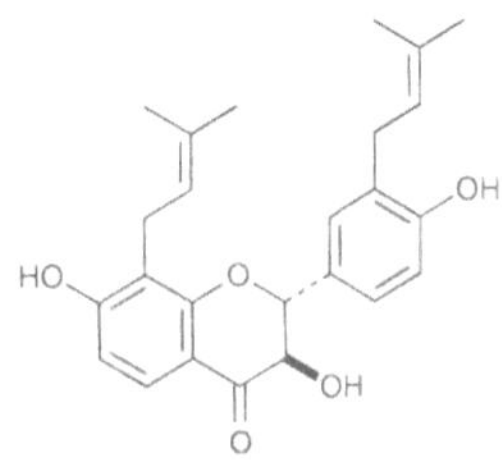

3-hydroxyglabrol (135)

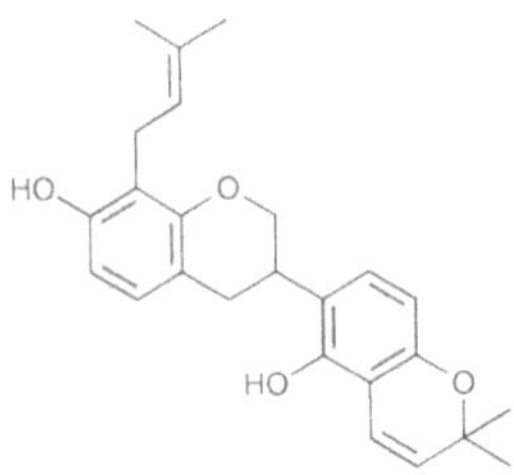

8-prenyl-phaseollin-isoflavan (136)

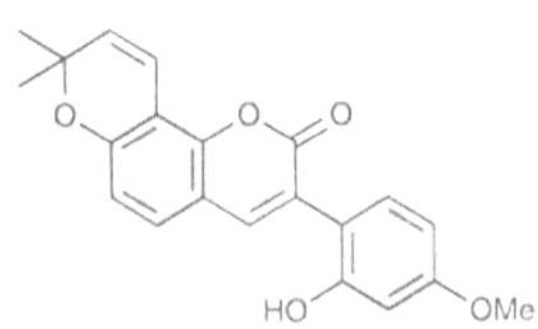

RL-P (137)

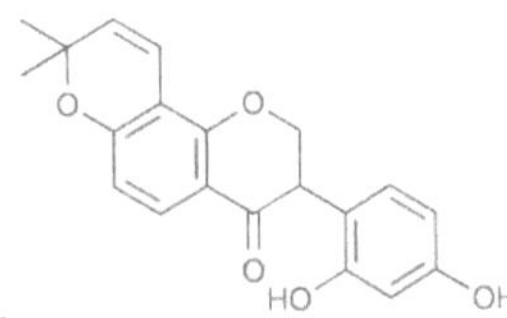

RL-Q (138)

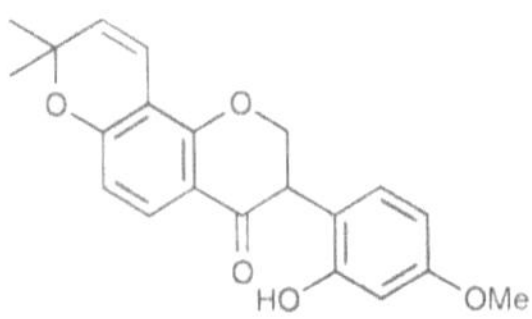

RL-R (139)

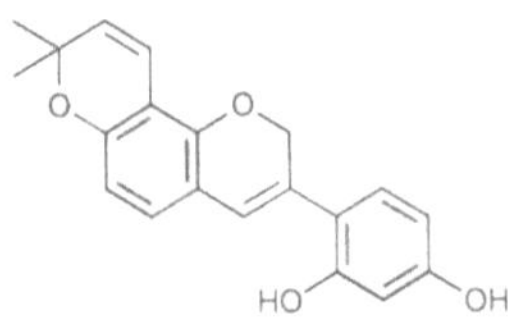

RL-S (140)

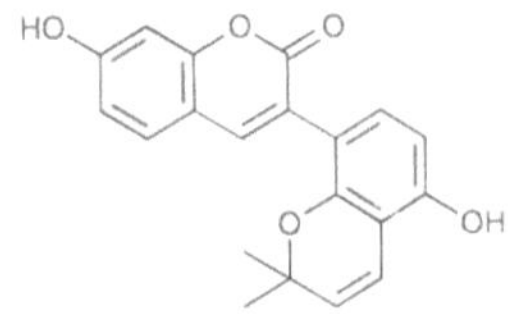

RL-U (141)

Fig. 10

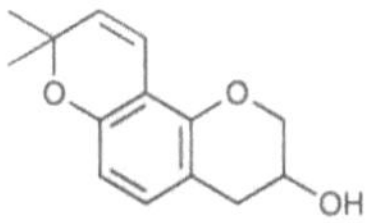

RL-V (**142**)

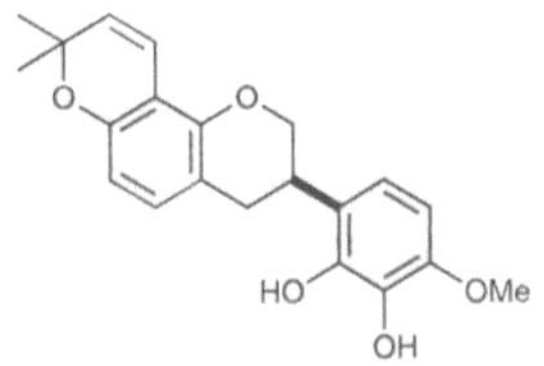

3'-hydroxy-4'-*O*-methyl-glabridin (**143**)

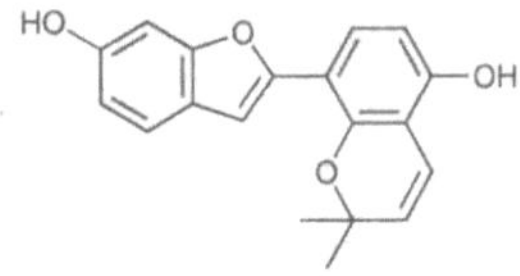

kanzonol U (**144**)

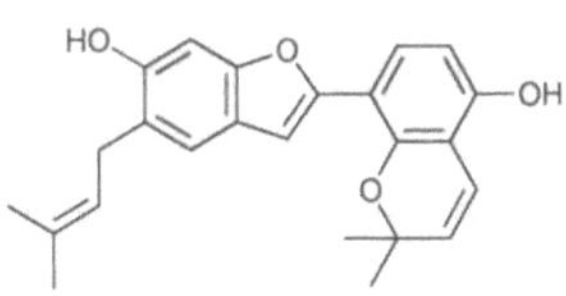

kanzonol V (**145**)

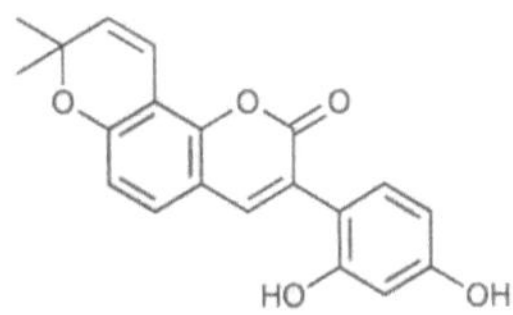

kanzonol W (**146**)

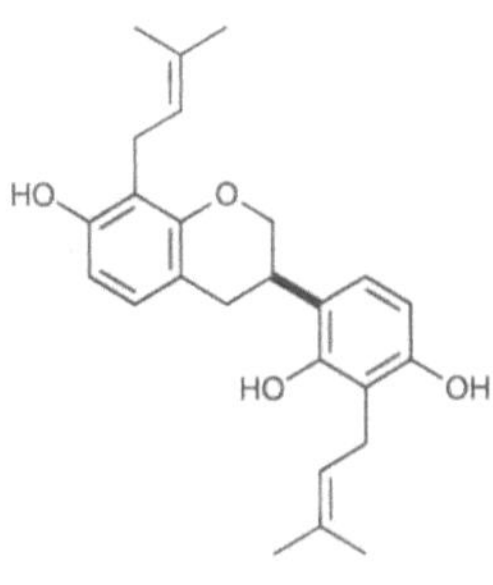

kanzonol X (**147**)

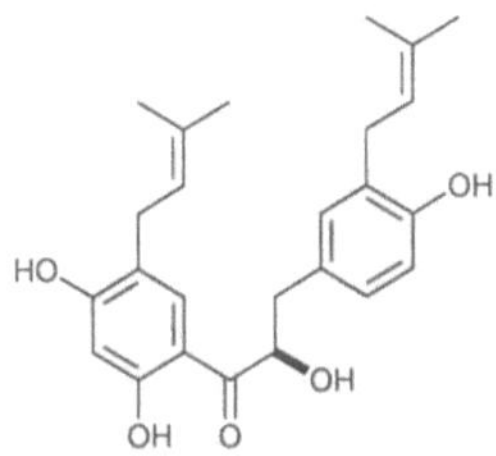

kanzonol Y (**148**)

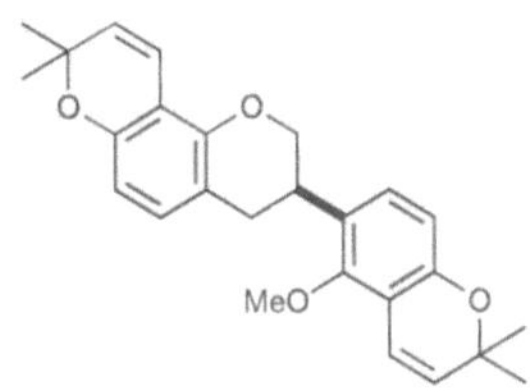

methylhispaglabridin B (**150**)

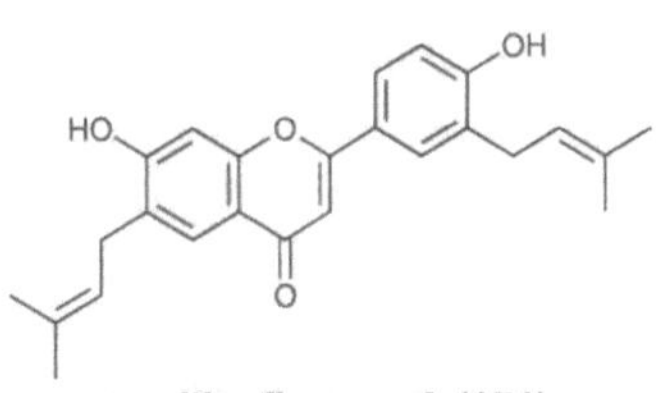

prenyllicoflavone A (**151**)
(**licoflavone B**)

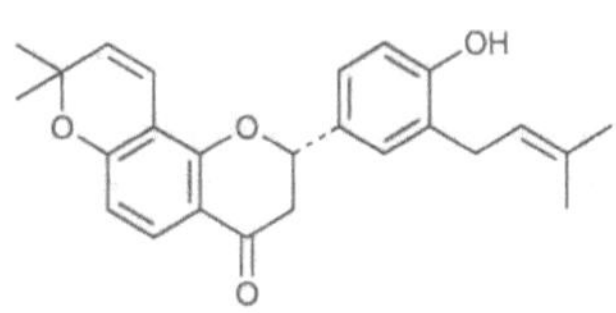

shinflavanone (**152**)

Fig. 11

isoflavan kanzonol X (**147**) and a diprenylated α-hydroxydihydrochalcone, kanzonol Y (**148**), together with some known compounds (*252*).

The isoflavans obtained from *G. glabra* are unusual in having no substituent at the C-5 position, while almost all isoflavans isolated from the other licorice species, *G. uralensis, G. aspera, etc.*, do as mentioned in later subsections.

KITAGAWA *et al.* have reported a new glucoside of liquiritin, glucoliquiritin apioside (**149**, Fig. 9), a pyranoisoflavan, methylhispaglabridin B (**150**), a diprenylated flavone, prenyllicoflavone A (**151**), a prenylated pyranoflavanone, shinflavanone (**152**), and two pyranopterocarpans, shinpterocarpin (**153**) and 1-methoxyphaseollin (**154**), from *G. glabra* collected in the Xinjiang Uigur Autonomous Region of China (in the extreme west of northwestern China) (*260*). The main phenolic constituents were licuraside (**119**, 1.65%) and liquiritin apioside (**115**, 0.98%). We isolated a new prenylated isoflavan, kanzonol R (**155**), and an isoprenoid-substituted pyranoisoflavone, kanzonol T (**156**), as minor components of material purchased in a market in Xinjiang (*261*, *263*). Ten known flavonoids, including licochalcone A, were also obtained from the Xinjiang licorice (*262*, *263*), but these compounds were not found in Spanish or Russian licorice.*

Of the above phenolic compounds obtained from *G. glabra*, compounds **109–115**, **119**, **120**, **126**, **128** and **151** have been also isolated from other *Glycyrrhiza* species (Table 13).

3.1.2. Glycyrrhiza uralensis

About sixty phenolic compounds have been isolated from the underground parts of this species. The main phenolic compounds are liquiritin (**110**, 0.6–3.7%) and isoliquiritin (**109**, 0.4–2.3%) (*36*). Four new glycosides, liquiritigenin 4′,7-diglucoside (**157**), 6″-acetylliquiritin (**158**), a feruloyl ester of licurazid (**159**, Fig. 9) and isoononin (**160**), have also been isolated from *G. uralensis* (*178*, *264–266*).

A diprenylated isoflavan (licoricidin) and three coumestans containing an isoprenoid group (glycyrol, 5-*O*-methylglycyrol [1-*O*-methylglycyrol in the new numbering system] and isoglycyrol) along with a prenylated isoflavone (licoricone) have been isolated from *G. uralensis* (*267–269*).† The structure of licoricone was established as **162** by X-ray

* Licochalcone A (**213**) is a prominent phenolic constituent of *G. inflata* and *G. eurycarpa* (subsections 3.1.4 and 3.1.5), and was isolated only from the *G. glabra* collections originating in Xinjiang (Table 13). The compound has been detected by HPLC analysis in some licorice collections from northwestern China, especially those from the Xinjiang Uigur Autonomous Region (*36*, *379*, *380*). In our more recent study of *G. glabra* imported from Kirghizia, the phenolic constituents of the licorice differ from those of Spanish and Russian licorice except for glabrene (**130**), and the constituents of Kirghiz licorice resemble to those of *G. glabra* in Xinjiang except for licochalcone A (**213**) (unpublished data).

† The species is mentioned in reviews (*74*, *78*) but not in the original articles (*267*, *268*).

analysis of its monobromoacetate (*270*). Syntheses of licoricone and related compounds have been carried out by TSUKAYAMA *et al.* (*271*). Formula **163** originally attributed to licoricidin on the basis of the ^{1}H NMR spectrum and the shift of an aromatic proton which accompanied acetylation (*267*) was revised to **164** using ^{13}C NMR spectrometry (*272*). The 7-*O*-methyl ether (**165**) of licoricidin was isolated from the roots and called licorisoflavan A (*273*, *274*). Synthesis confirmed the structure but left unassigned stereochemistry at C-3 (*275*).

Formula **166** was originally proposed for glycyrol because of the ^{1}H NMR data, the absence of a shift in the UV spectrum induced by sodium acetate in ethanol supposedly indicative of the absence of a free hydroxyl group on C-3 and a positive Gibbs test (*268*). The structure of isoglycyrol which is formed from glycyrol under acidic conditions was therefore assumed to be **167**. In our experience the diagnostic bathochromic shift induced by sodium acetate in methanol is not paralleled in ethanol and some coumarins and flavonols with a substituent *para* to a hydroxyl group do react with the Gibbs reagent to give a blue or green pigment (*276*). X-ray analysis of isoglycyrol led to the revised structure **168** from which it followed that glycyrol must have structure **169** (*277*). More or less simultaneously KINOSHITA *et al.* (*278*) came to the same conclusion on the basis of NOE studies on the dibenzyl ether of glycyrol. Compound (**161**) is therefore not 1-*O*-methylglycyrol, but 3-*O*-methylglycyrol.

ZHU *et al.* have isolated a prenylated 2-arylbenzofuran with a C-3 hydroxyl group, liconeolignan (**170**) (an earlier tentative name was licobenzofuran), together with a prenylflavonol, isolicoflavonol (**171**), and a prenylated 3-arylcoumarin, glycycoumarin (**172**), from the roots and stems of *G. uralensis* (*279–281*). Two isoflavones, semilicoisoflavone B (**173**) and isoangustone A (**174**), a deoxybenzoin derivative, licoriphenone (**175**), and a diprenylated pterocarpan, 1-methoxyficifolinol (**176**), have been isolated from commercially available licorice (*G. uralensis*, although the material was not unequivocally identified) by TSUDA *et al.* (*282*). Licoriphenone (**175**) is converted to a 2-arylbenzofuran (gancaonin I, **177**) with polyphosphoric acid in dichloromethane (*282*). A dibenzyl ether of **175** has been used as a starting material in the synthesis of a 2-arylbenzofuran derivative (*283*).

In our study of the species, the main phenolic compounds with prenyl group(s) were licoricidin (**164**, 0.011%), licorisoflavan A (**165**, 0.0025%) and glycyrol (**169**, 0.0023%). We also isolated twelve new isoprenoid-substituted phenols as minor constituents, kanzonols F–Q (**178–189**), from the underground parts (*261*, *284*, *285*). Kanzonols M (**185**), N (**186**), and O (**187**) are rare isoflavans with a formyl group on the A ring. The structures of the diprenylated isoflavones, kanzonols K (**183**) and L

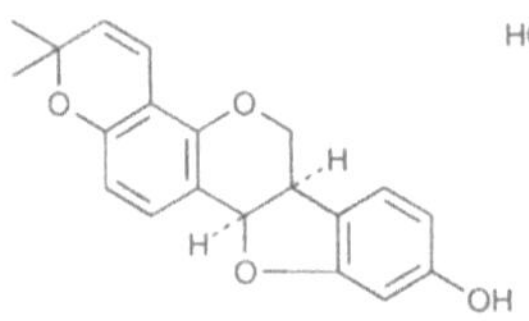

shinpterocarpin (153)

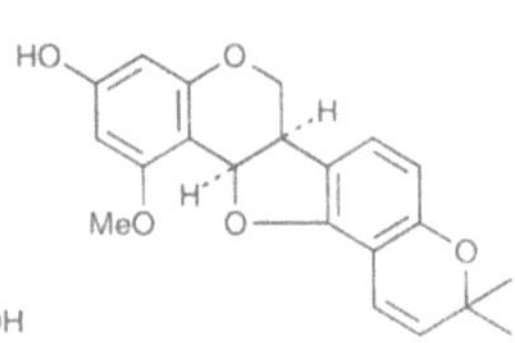

1-methoxyphaseollin (154)

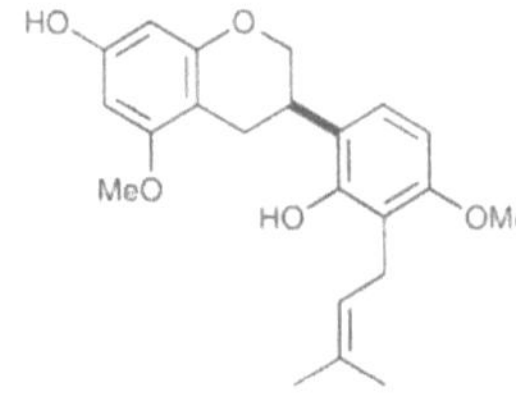

kanzonol R (155)

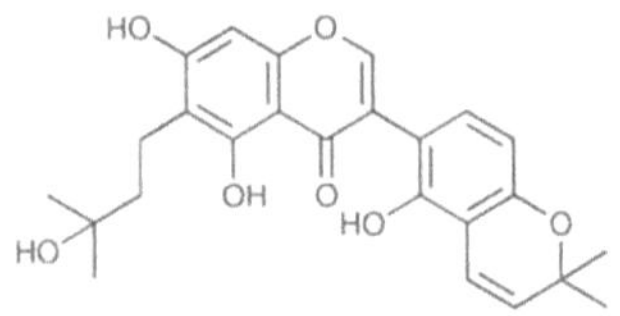

kanzonol T (156)

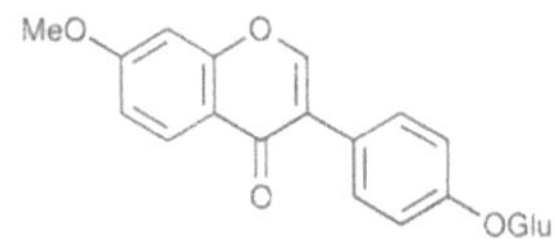

isoononin (160)

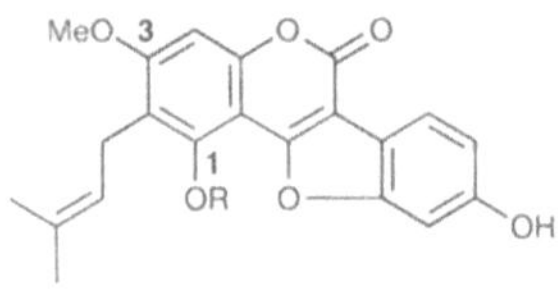

3-*O*-methylglycyrol (161):
R = Me
(**166**): R = H

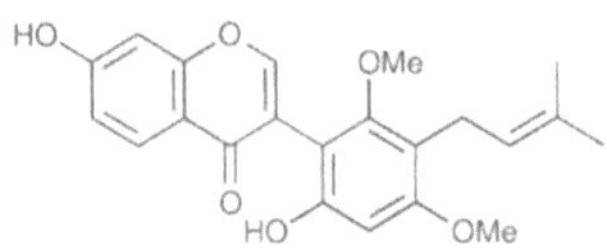

licoricone (162)

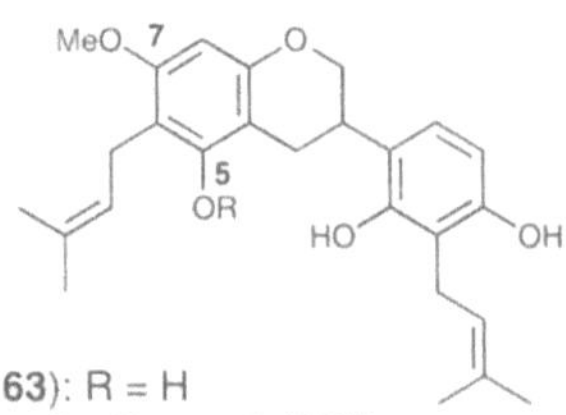

(**163**): R = H
licorisoflavan A (165):
R = Me

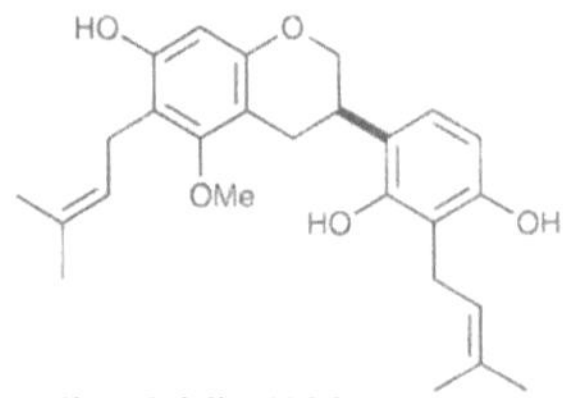

licoricidin (164)

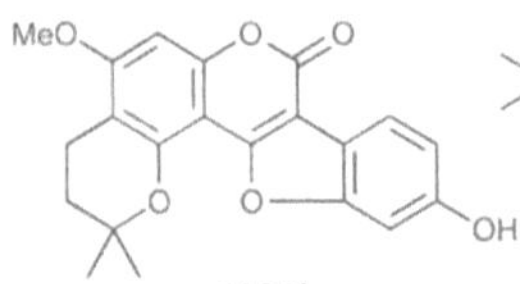

(**167**)

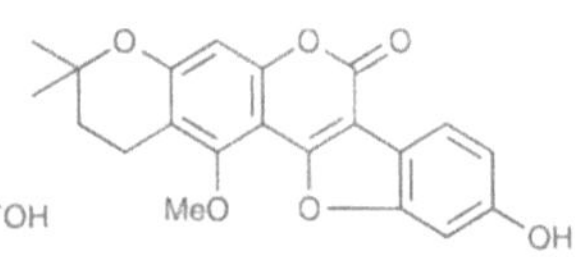

isoglycyrol (168)

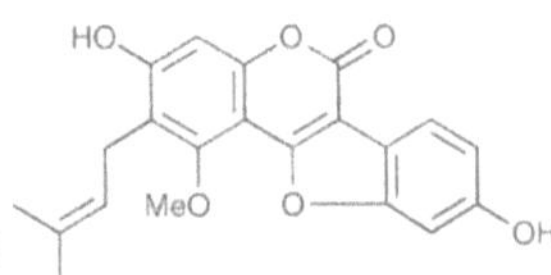

glycyrol (169)

Fig. 12

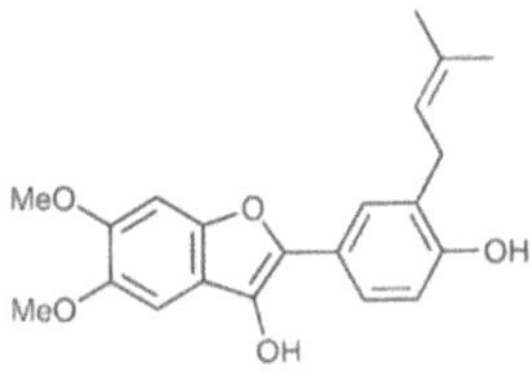

liconeolignan (170)

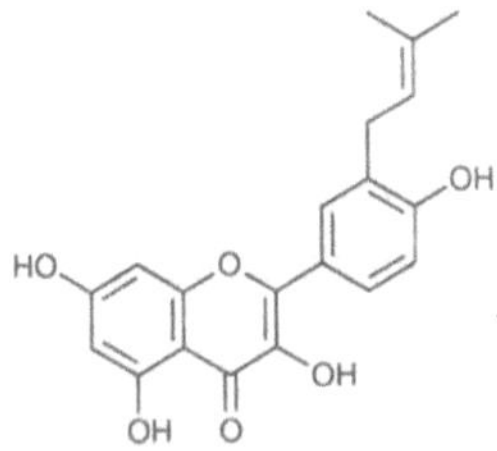

isolicoflavonol (171)

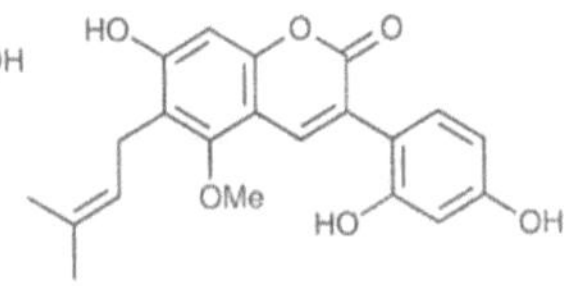

glycycoumarin (172)

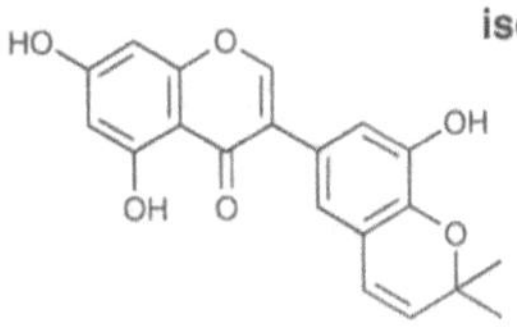

semilicoisoflavone B (173)

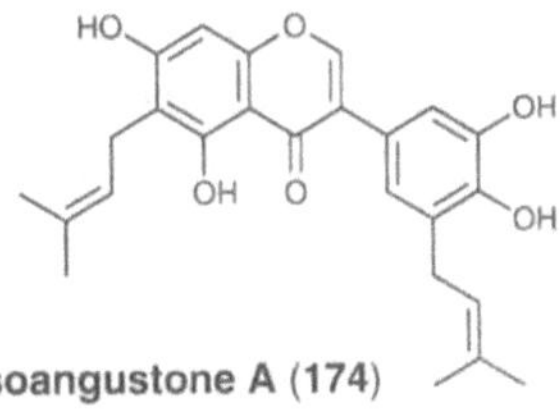

isoangustone A (174)

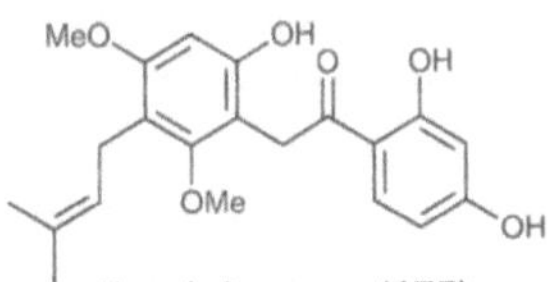

licoriphenone (175)

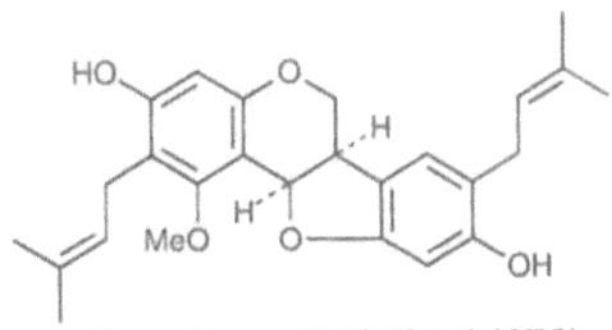

1-methoxyficifolinol (176)

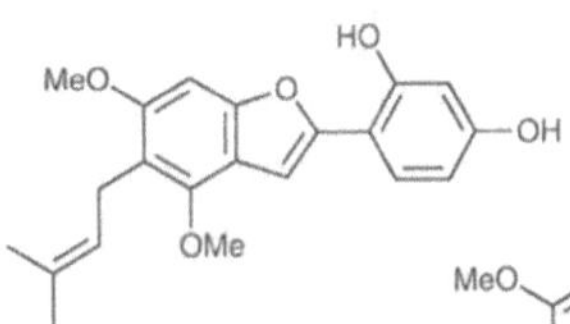

gancaonin I (177)

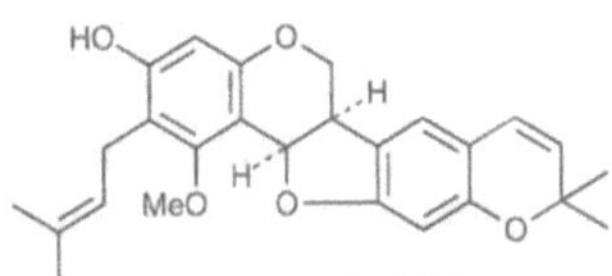

kanzonol F (178)

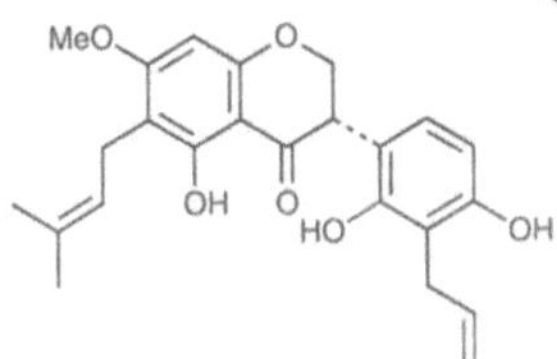

kanzonol G (179)

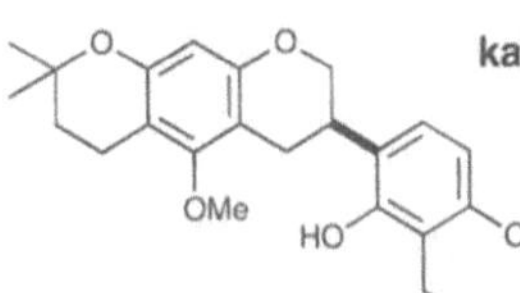

kanzonol H (180)

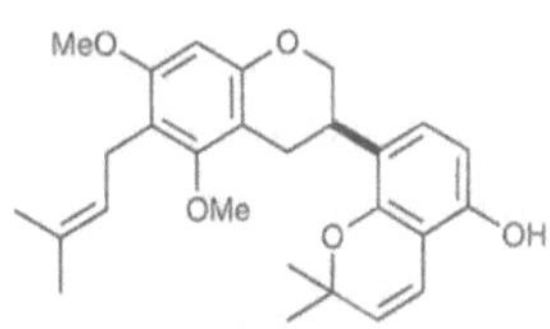

kanzonol I (181)

Fig. 13

(**184**), were elucidated using as criterion the chemical shift of the proton of the hydrogen-bonded hydroxyl group because their yields were very low (*285*). This new method for the structure determination is described in Section 4. Recently a chalcone (**190**) and a 3-arylcoumarin (**191**) have also been reported from *G. uralensis* (*286*).

3.1.3. Glycyrrhiza aspera

About forty phenolic compounds have been isolated from the underground parts of *G. aspera*. The main phenolic compounds are liquiritin (**110**, 0.33%) and isoliquiritin (**109**, 0.20%), and the main phenols with isoprenoid substituent(s) are the isoflavonoids licoricidin (**164**, 0.056%), licorisoflavan A (**165**, 0.013%) and licoisoflavanone (**192**, 0.015%), the latter having been isolated earlier as a triacetate from an unidentified Xinjiang licorice (*287*).

Fourteen new isoprenoid-substituted phenols, glyasperins A–N (**193**–**206**), were found as minor compounds together with some flavonoids known from *G. aspera* collected in the Xinjiang Uigur Autonomous Region (*288–291*).* The prenylated isoflavan (±)-glyasperin D (**196**) had been synthesized previously by Shih *et al.* (*275*). Except for glyasperin H (**200**) the 5-position of these flavonoids is substituted with a hydroxyl or methoxyl group. Among them, four compounds, glyasperin A (**193**), kumatakenin (4′,5-dihydroxy-3,7-dimethoxyflavone), topazolin (**207**), licoflavonol (**208**), are flavonols, one is a 2-arylbenzofuran (licocoumarone, **209**), and one is a chromen-4-one derivative (**197**, glyasperin E), but the others are isoflavonoids. Glyasperin E (**197**) is a 3-oxygenated chromen-4-one derivative with a phenoxyl group and a prenyl group. Although 3-phenoxychromen-4-one derivatives have been synthesized (*292*), glyasperin E is the first example of natural product with such a skeleton. The structure was elucidated by spectroscopic methods and confirmed by synthesis of glyasperin E dimethyl ether (**197a**) (*289*).

A 3-arylcoumarin, isoglycycoumarin (**210**), has been isolated from Chinese licorice by Zhang *et al.* (*98*). A prenylated isoflav-3-ene dehydroglyasperin C (**211**) and a *O*-prenylated pterocarpan (**212**) have been isolated from material collected in Xinjiang by Kitagawa *et al.* (*293*).

---

* For our studies of the licorice grown in northwestern China, the material was collected by Dr. L. Zeng in a particular location where only one kind of licorice grows. Generally different species of licorice whose roots and stolons intermingle grow in this area, thus making it difficult to obtain material representing a single species.

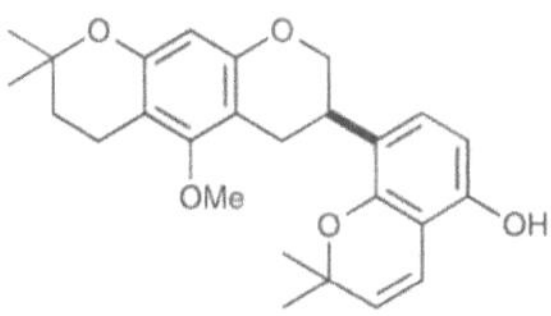

kanzonol J (182)

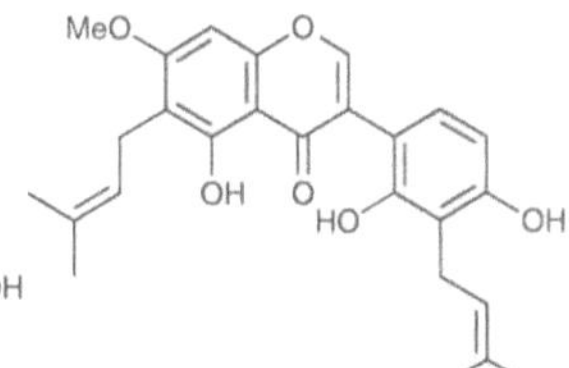

kanzonol K (183)

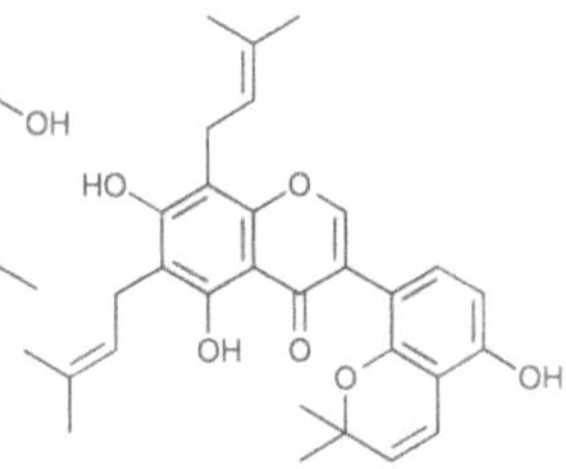

kanzonol L (184)

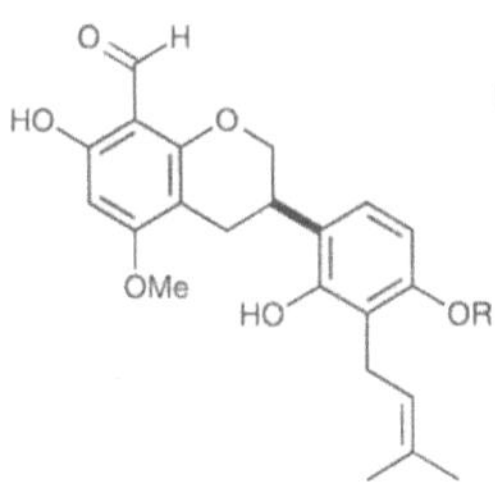

kanzonol M (185): R = Me
kanzonol N (186): R = H

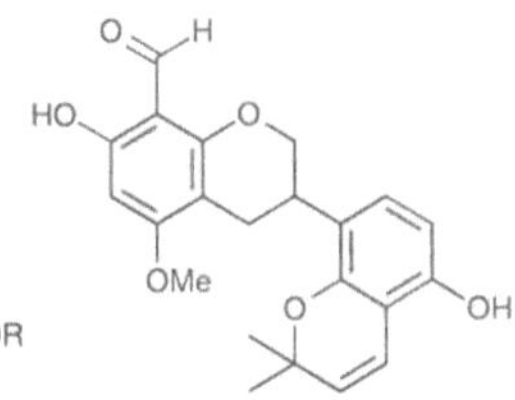

kanzonol O (187)

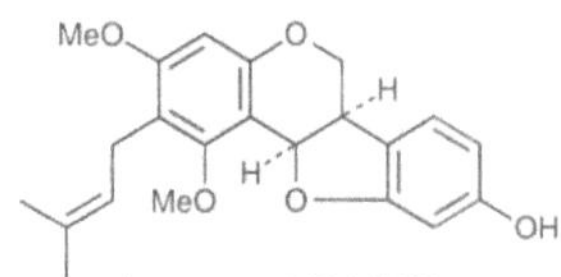

kanzonol P (188)

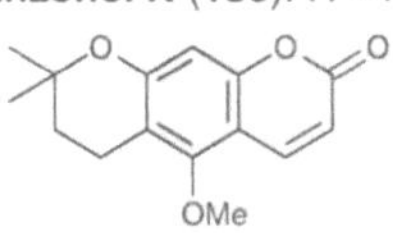

kanzonol Q (189)
(dihydroxanthoxyletin)

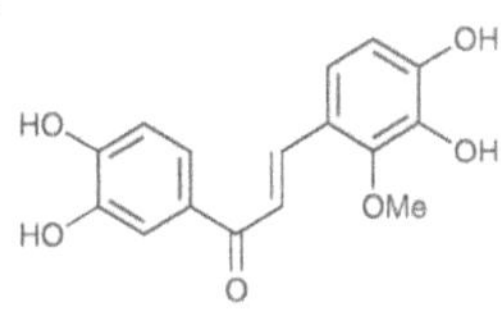

(190)

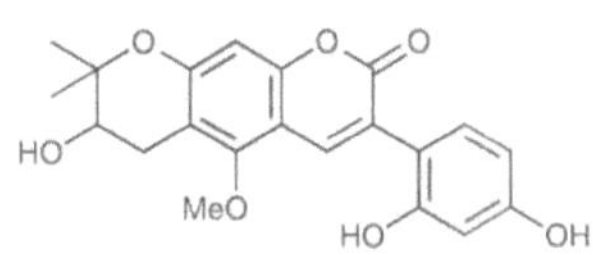

(191)

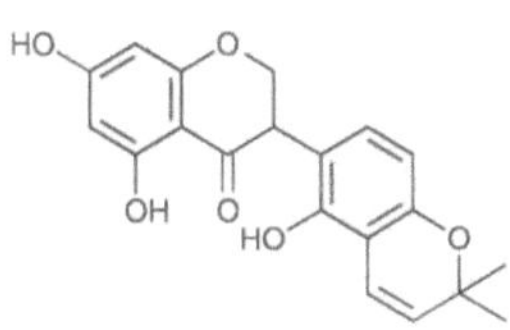

licoisoflavanone (192)

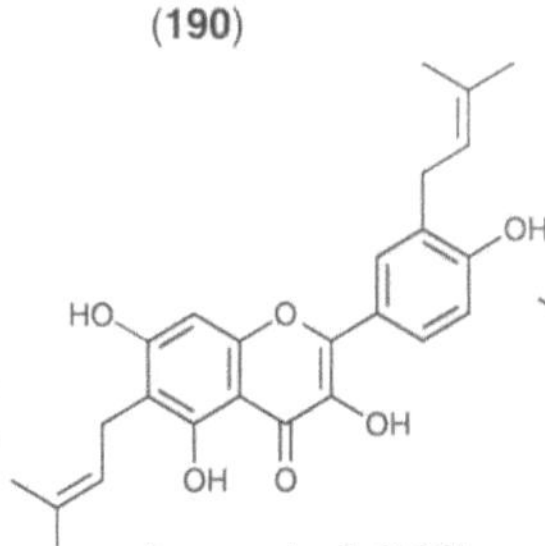

glyasperin A (193)

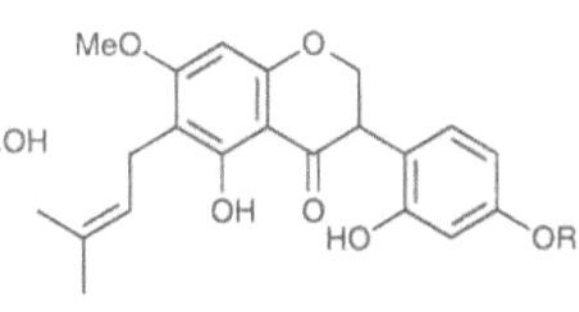

glyasperin B (194): R = H
glyasperin K (203): R = Me

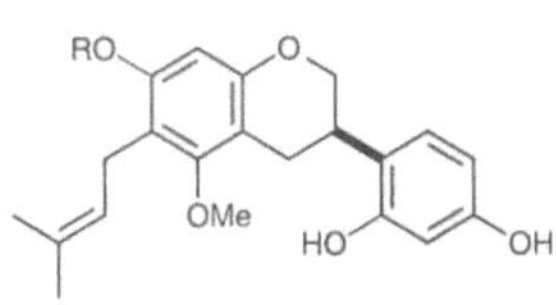

glyasperin C (195): R = H
glyasperin D (196): R = Me

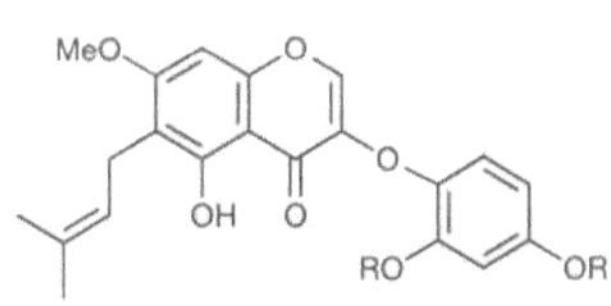

glyasperin E (197): R = H
(197a): R = Me

Fig. 14

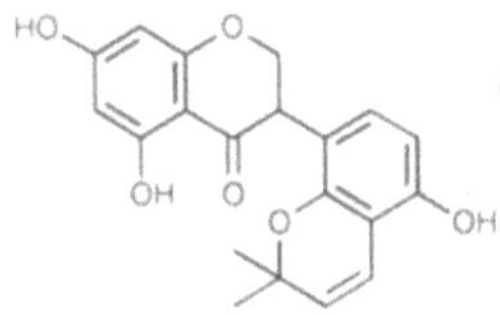

glyasperin F (198)

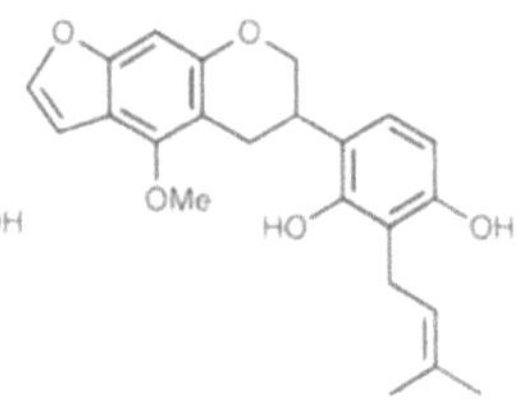

glyasperin G (199)

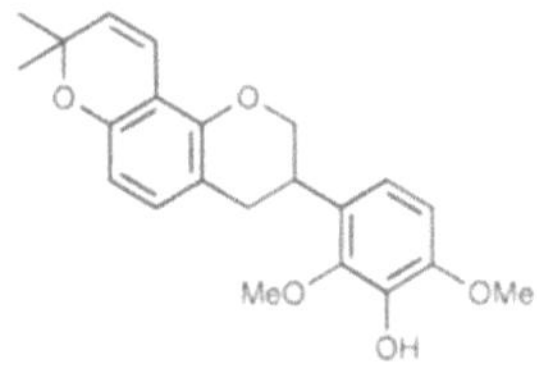

glyasperin H (200)

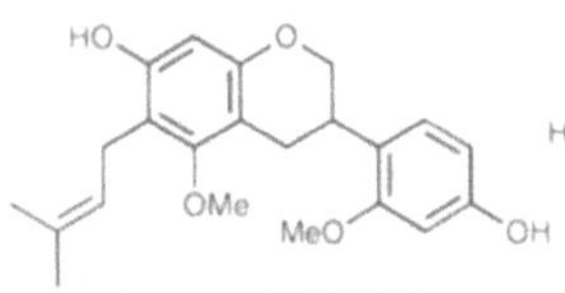

glyasperin I (201)

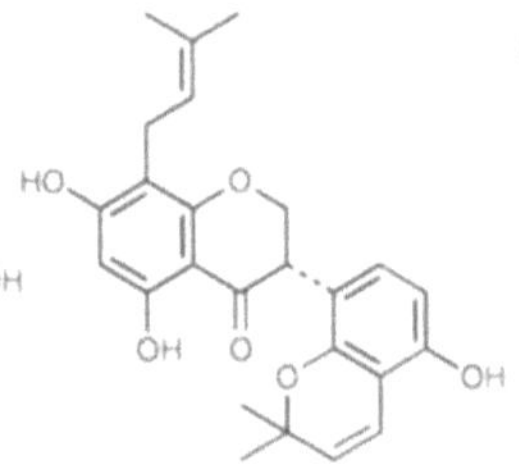

glyasperin J (202)

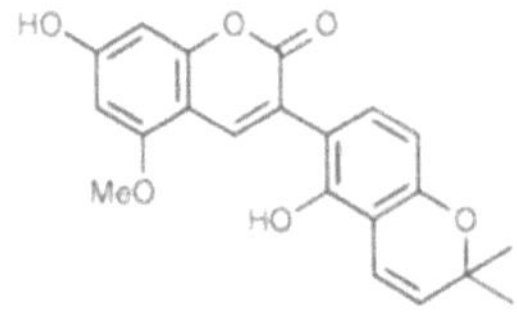

glyasperin L (204)

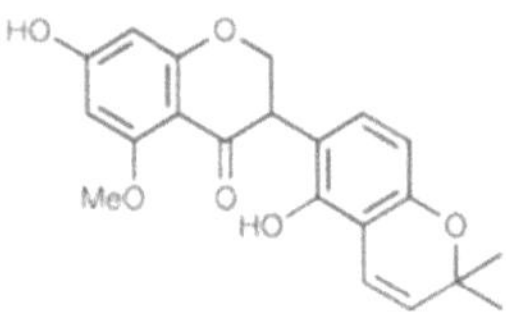

glyasperin M (205)

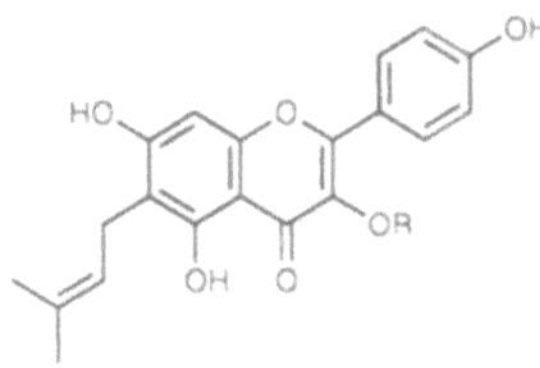

topazolin (207): R = Me
licoflavonol (208): R = H

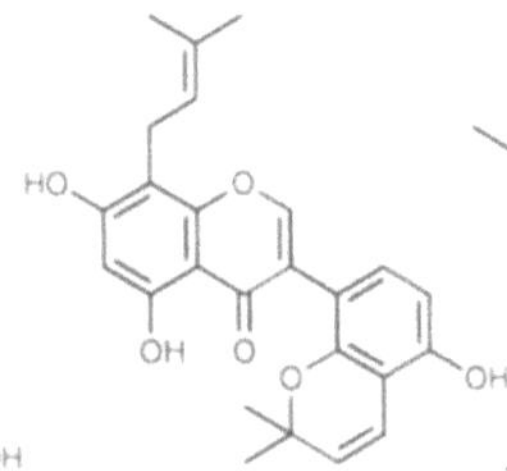

glyasperin N (206)

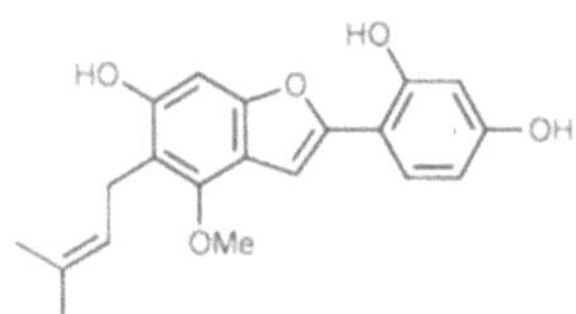

licocoumarone (209)

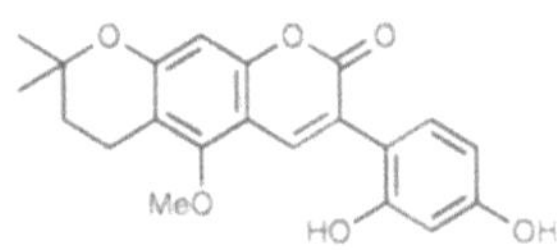

isoglycycoumarin (210)

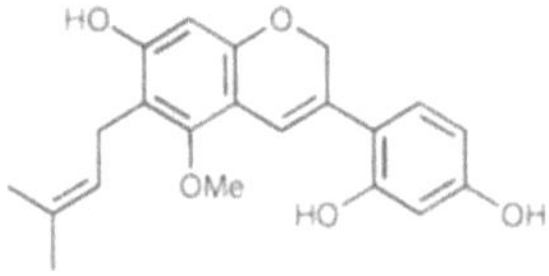

dehydroglyasperin C (211)

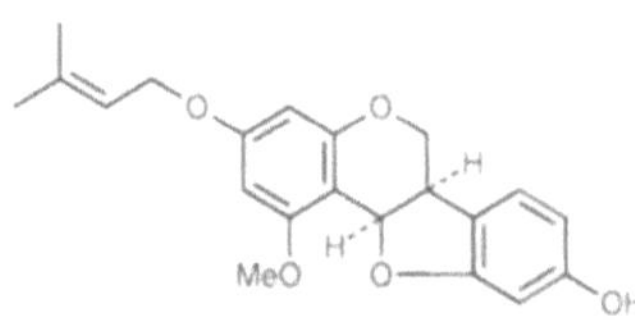

(212)

Fig. 15

3.1.4. Glycyrrhiza inflata

This licorice is the main species in Xinjiang licorice and is used as a crude drug in China (*79*). In our study of a commercially available Xinjiang licorice, the crude drug consisted of *G. inflata* (50%), *G. eurycarpa* (7.5%), *G. uralensis* (2.5%) and some unidentified material (40%) that may represent hybrids (*294*). About sixty phenolic compounds have been isolated. The main phenols in the roots of *G. inflata* are liquiritin (**110**, 0.2–0.6%) and isoliqiritin (**109**, 0.1–0.5%) (*36*), and the main isoprenoid-substituted flavonoids are licochalcone A (**213**, 0.12%) and glabrone (**128**, 0.065%) (*295*).

Saitoh and Shibata have isolated licochalcones A and B from Xinjiang licorice and elucidated their structures as **213** and **214**, respectively, using ^{1}H NMR and mass spectrometric data (*296*). They applied the name retrochalcone to this type of chalcone with a methoxyl group at C-2 and no substituent at C-2′ and C-6′ (*287*, *297*).* This type of chalcone was proved to be produced biosynthetically by a route reverse to the one by which ordinary chalcones are formed; the A ring is derived from shikimate and the B ring from polyketide of malonate origin (*296*) (biosynthesic studies of this type chalcone echinatin will be described in a following section). At the same time Chinese researchers isolated licochalcone A from *G. inflata* and established its structure by synthesis (*298*). A pyranoisoflavone, licoisoflavone B (**215**), has also been isolated as triacetate along with licoisoflavanone (**192**) from Xinjiang licorice by Shibata *et al.* (*287*).

We examined the phenolic constituents of the roots of *G. inflata* collected in Gansu Province in northwestern China (*295*, *299*). Thirty phenolic compounds were obtained eleven of which were new, glyinflanins A–K (**216**–**226**). The ^{1}H and ^{13}C NMR spectra of the dibenzoylmethanes, glyinflanins A–F (**216**–**221**), indicate the presence of an equilibrium mixture of tautomeric dibenzoylmethanes and β-hydroxychalcones. Evidence for the presence of an equilibrium mixture is easily obtained by deuterium exchange of the α-protons of the tautomers. Thus, the ^{1}H NMR spectrum of **216** in acetone-d_6 exhibits singlets at δ 4.54 (0.4H × 2) and 6.85 (0.6H) which disappeared immediately after addition of CD_3OD and can be attributed to the C-2-H (α-protons of **216a**) of dibenzoylmethane existing in the equilibrium

* Recently Dr. Shibata has informed us that he wishes to change the name "retro-chalcone" to "reversely constructed chalcone" based precisely on the biosynthetical scheme.

mixture of the keto form (**216**) and the enol form (**216a**). In the solid state, these compounds may exist in the dibenzoylmethane form because the ratio of two tautomers depends upon the concentration of solution; in 10% (w/v) solution of **216**, the ratio of keto and enol forms is about 1:1, and in a 2% solution of **216**, the ratio changes to about 2:3 (*295*). The β-hydroxychalcone (**216a**) has two hydrogen-bonded hydroxyl groups, with the hydrogen bond between the β-hydroxy proton and the carbonyl oxygen being stronger (δ 15.78) than the other hydrogen bond (δ 12.27). The hydrogen bond (between the 2′-OH and the carbonyl oxygen) is weaker than the intramolecular hydrogen bond of its tautomer (**216**). Dibenzoylmethanes are well known in flavonoid chemistry as intermediates in flavone synthesis or in the Wessely-Moser rearrangement (*300, 301*). Glyinflanin A (**216**) is converted to prenyllicoflavone A (**151**, Fig. 11) by heating in dry benzene (*295*).

The other new compounds are a chalcone having two 2,2-dimethylpyran rings, glyinflanin G (**222**), a pyrano-2-arylbenzofuran glyinflanin H (**223**) and three pyranoisoflavans glyinflanins I–K (**224**–**226**). A prenylflavone, licoflavone C (**227**),* and a dibenzoylmethane, glycyrdione B (**228**), have also been isolated together with two reversely constructed chalcones, licochalcones C (**229**) and D (**230**), by KAJIYAMA *et al.* (*302–304*). A prenylated pyranochalcone (**231**) and a neoflavone, inflacoumarin A (**232**), have also been reported (*305–307*).

3.1.5. Glycyrrhiza eurycarpa

This licorice species grows in northwestern China as described in an earlier section and was earlier assigned to *G. korshinskyi* (*34*). Li examined the material and showed that it is a hybrid between *G. uralensis* and *G. inflata* (*33*), but has assigned it species rank because it is a naturally occurring hybrid and named it *G. eurycarpa* P.C. Li.

About forty phenolic compounds have been isolated from the underground parts of this species. The main phenolic compounds are liquiritin (**110**, 0.8–1.3%) and isoliquiritin (**109**, 0.6–0.7%) as in the other Chinese species (*36*). LIU *et al.* have isolated a glycoside of formononetin, glycyroside (**233**), a 3-hydroxy-3-methylglutaroyl ester of

* The trivial name licoflavone B was used for 6-prenylapigenin by KINOSHITA *et al.* (*381*). The same name was also used for prenyllicoflavone A (**151**) by KAJIYAMA *et al.* (*302*). The compound earlier assigned as 6-prenylapigenin and licoflavone C (8-prenylated apigenin) (**227**) may be same compound (private communication from Dr. KINOSHITA). Hence Dr. KINOSHITA has retracted his earlier name licoflavone B for 6-prenylapigenin, thus reserving the trivial name (licoflavone B) for **151**.

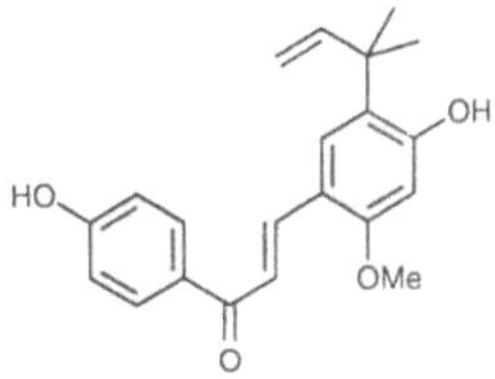

licochalcone A (213)

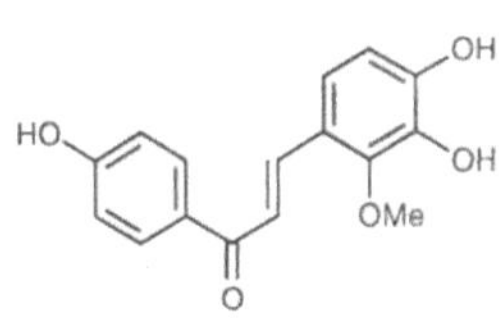

licochalcone B (214)

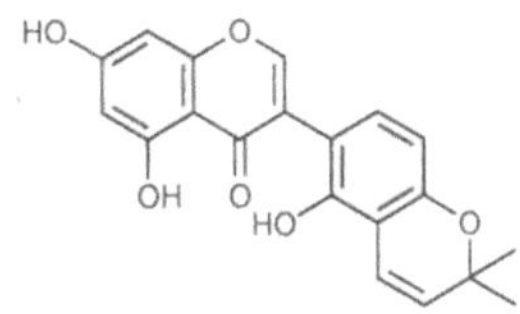

licoisoflavone B (215)

glyinflanin A (216)
(glycyrdione A)

(216a)

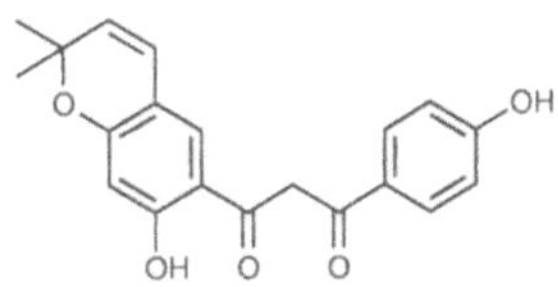

glyinflanin B (217)

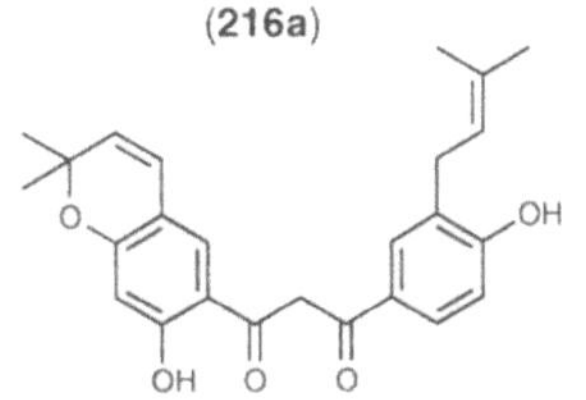

glyinflanin C (218)
(glycyrdione C)

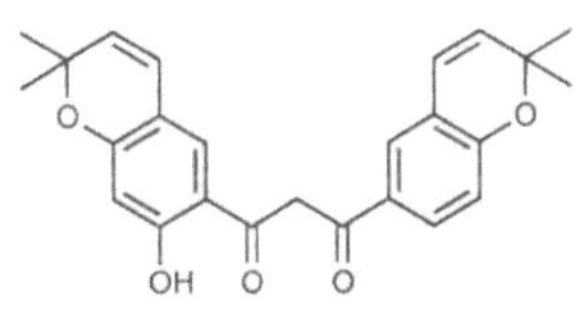

glyinflanin D (219)

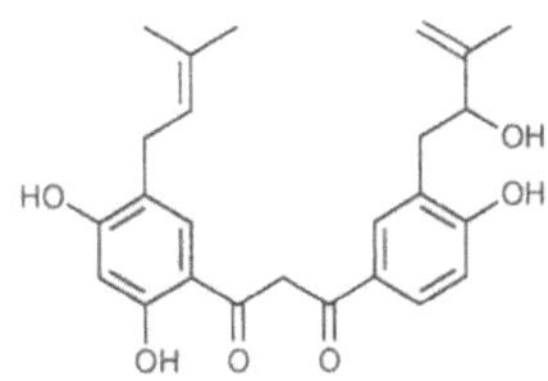

glyinflanin E (220)

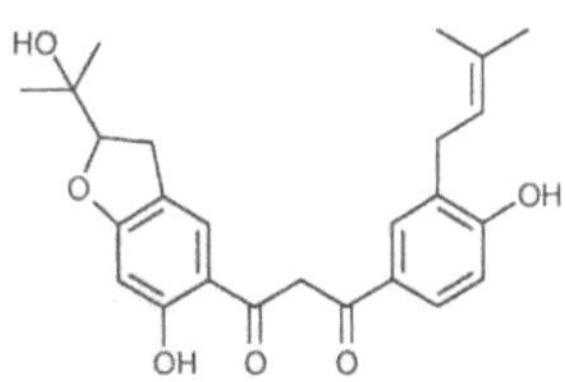

glyinflanin F (221)

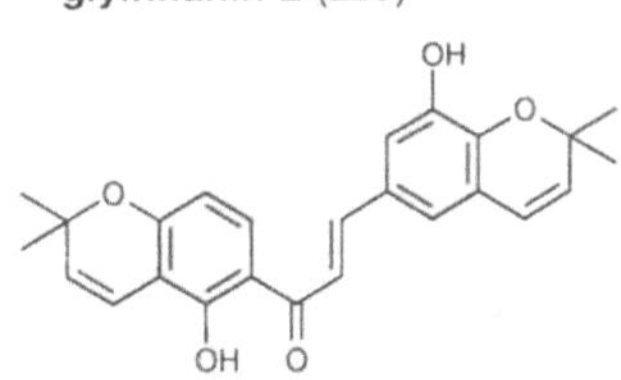

glyinflanin G (222)

Fig. 16

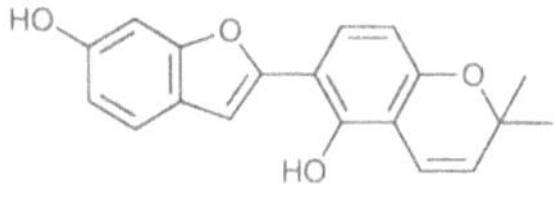

glyinflanin H (223)

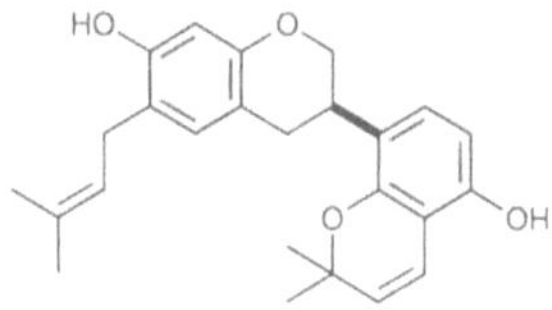

glyinflanin I (224)

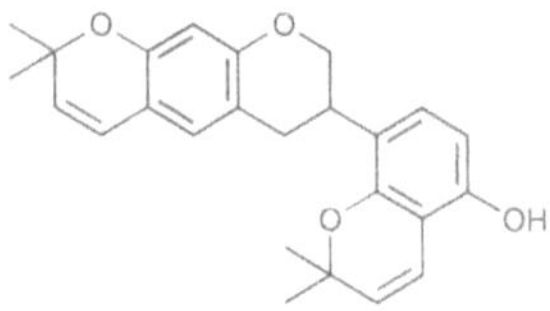

glyinflanin J (225)

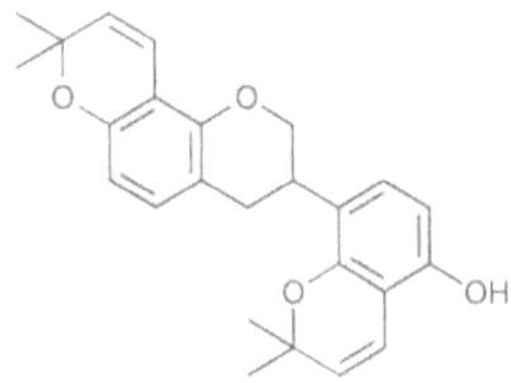

glyinflanin K (226)

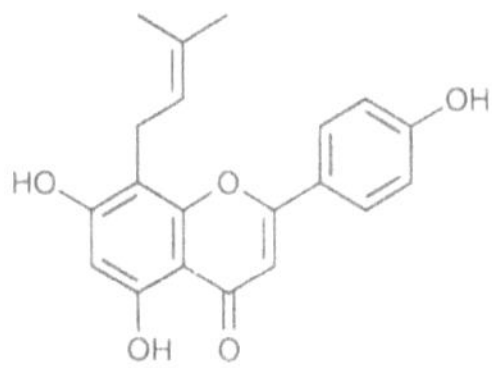

licoflavone C (227)

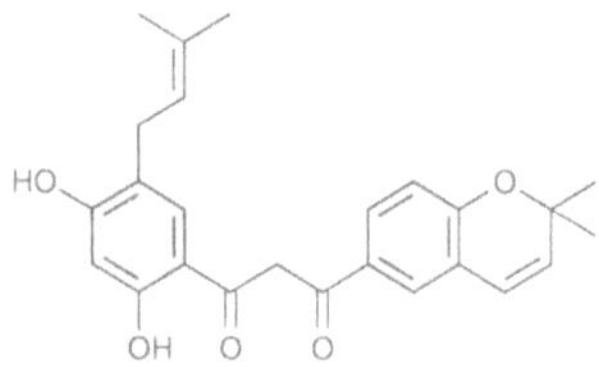

glycyrdione B (228)

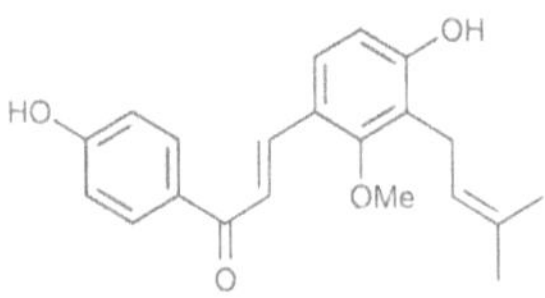

licochalcone C (229)

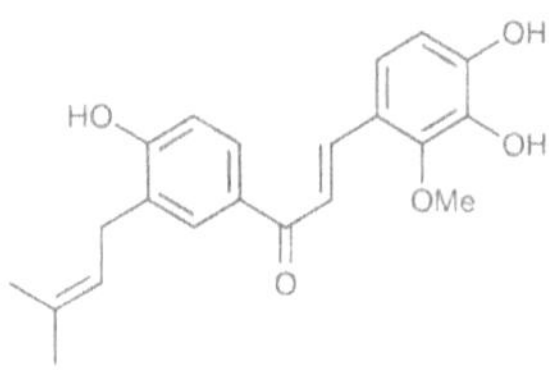

licochalcone D (230)

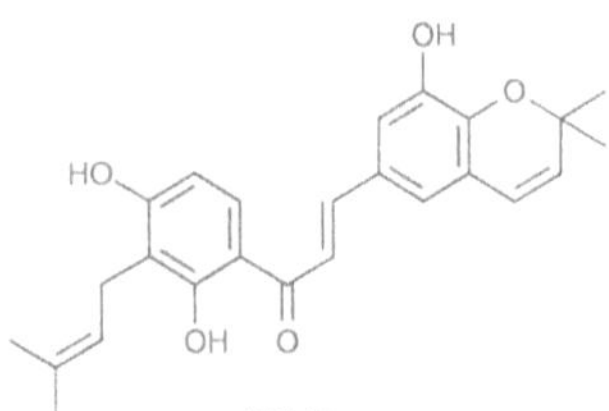

(231)

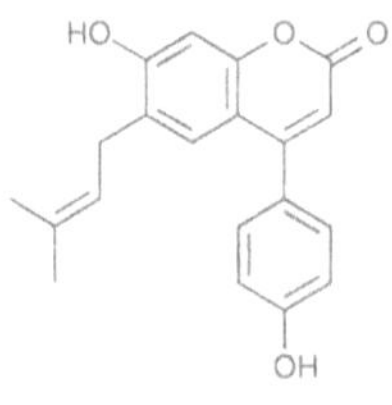

inflacoumarin A (232)

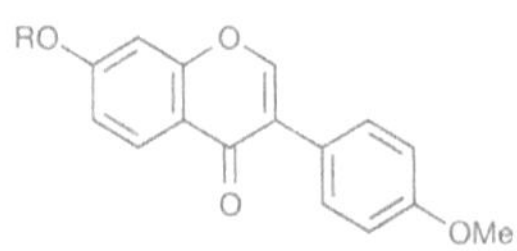

glycyroside (233):
R = Api(1 → 2)Glu-
formononetin (242):
R = H

Fig. 17

isoviolanthin (**234**), and a prenylated isoflavone, eurycarpin A (**235**) (*308–311*).

We examined the phenolic constituents of *G. eurycarpa* separated from commercial Xinjiang licorice mentioned in the preceding subsection and isolated eleven compounds (*294*). Among these, five are new compounds and were named kanzonols A–E (**236–240**). Kanzonol A (**236**) is a dibenzoylmethane having a prenyl group, kanzonols B (**237**) and C (**238**) are chalcones, and kanzonols D (**239**) and E (**240**) are prenylated flavones. The main isoprenoid-substituted phenol of this licorice is licochalcone A (**213**, 0.37%) as in *G. inflata*. Non-glycosidic phenols common to *G. uralensis*, *G. inflata* and *G. eurycarpa* are glycycoumarin (**172**), isoglycycoumarin (**210**), licoflavone A (**241**) (*74*), formononetin (**242**) and 4′,7-dihydroxyflavone (Table 13). Licochalcone A (**213**), glyinflanin B (**217**) and 5′-prenyllicodione (isolated earlier from the cultured cells of *G. echinata*) (**243**) have been isolated from both *G. inflata* and *G. eurycarpa*.

3.1.6. Glycyrrhiza pallidiflora

Thirty phenolic compounds have been isolated from the underground parts of this species. We isolated a diprenylated dihydrochalcone gancaonin J (**244**) and a new type of homoisoflavanone gancaonin K (**245**, in Scheme 1) from the roots of plants collected in northeastern of China (*312*). In consideration of the biogenesis of homoisoflavonoids (*313*), gancaonin K (**245**) might be derived from 2′-*O*-methyllicodione (**246**) obtained from the same source, as shown in Scheme 1. The dibenzoylmethane (**246**) has been synthesized earlier enzymatically (*314*) as the first example of such a synthesis of a compound occurring in licorice.

A new isoflav-3-ene, pallidiflorene (2′,7-dihydroxy-4′-methoxyiso-flav-3-ene), has been isolated from this species (*78*, *315*). A new isoflavone pallidiflorin (5-hydroxy-4′-methoxyisoflavone) was isolated by Liu *et al.* (*196*), and a new chalcone glypallichalcone (**247**) has been isolated from the roots by Zhang *et al.* (*316*).

3.1.7. Commercial Licorice in Japan (Northwest Licorice)

Many types of licorice have been imported into Japan. In the Japanese market, Chinese licorice is classified by its place of production, *e.g.*, Northeast licorice (Tohoku kanzoh in Japanese), Northwest licorice (Seihoku kanzoh), Xinjiang licorice (Shinkyo kanzoh), *etc.* (*317–325*). Among these Northeast licorice (*G. uralensis*) was the usual source for

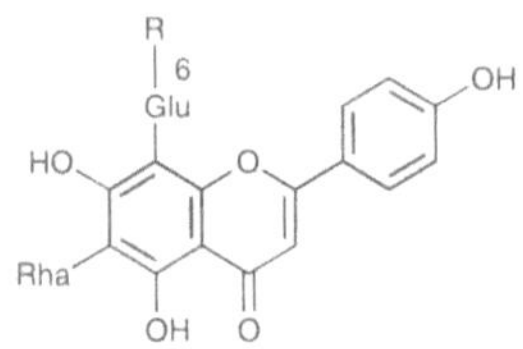

(234): R = $-C(=O)-CH_2-C(OH)(CH_3)-CH_2COOH$

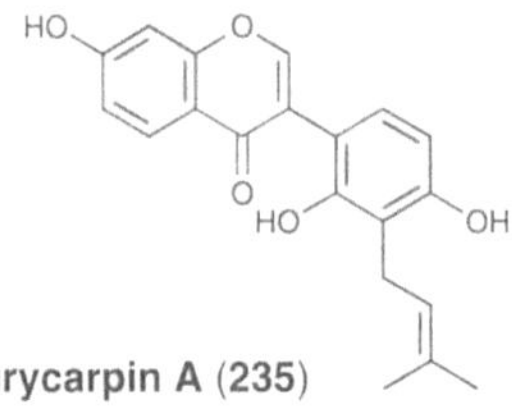

eurycarpin A (235)

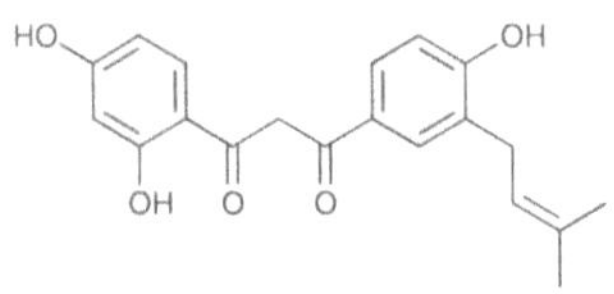

kanzonol A (236)

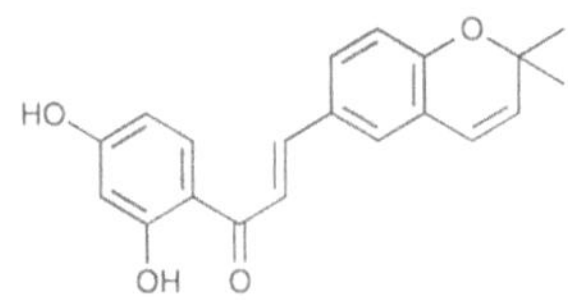

kanzonol B (237)

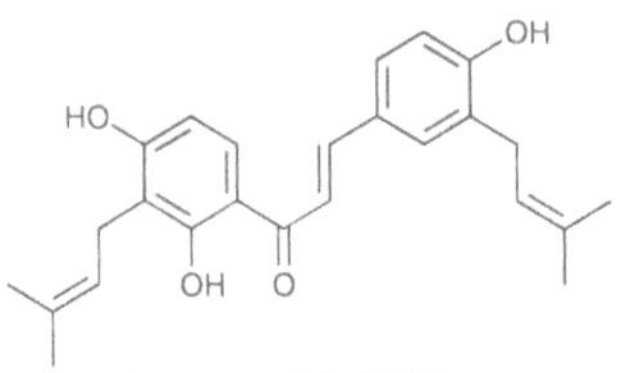

kanzonol C (238)

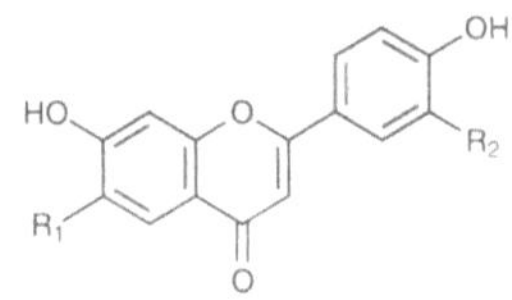

kanzonol D (239): R_1 = H, R_2 = prenyl
licoflavone A (241): R_1 = prenyl, R_2 = H

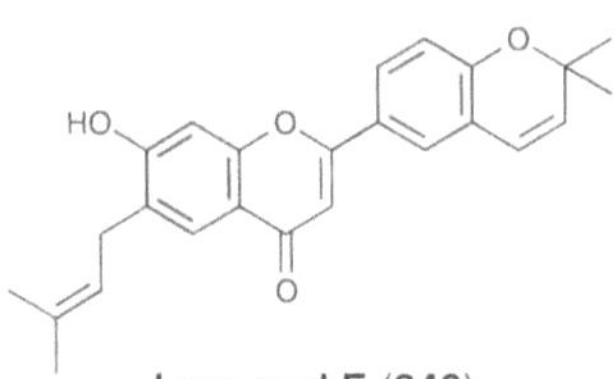

kanzonol E (240)

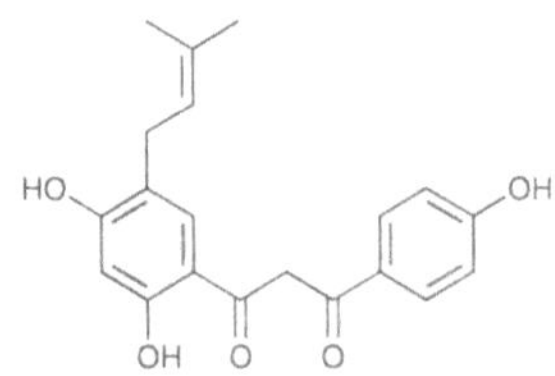

5'-prenyllicodione (243)

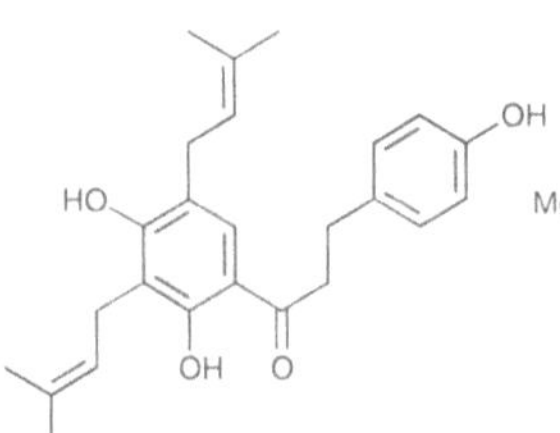

gancaonin J (244)

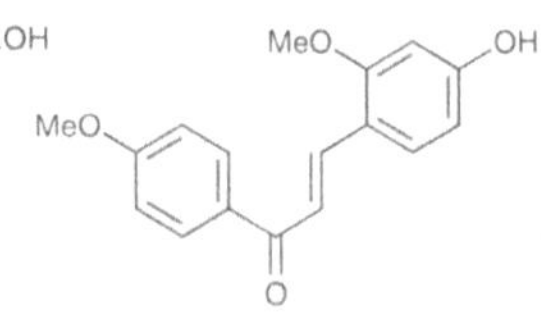

glypallichalcone (247)

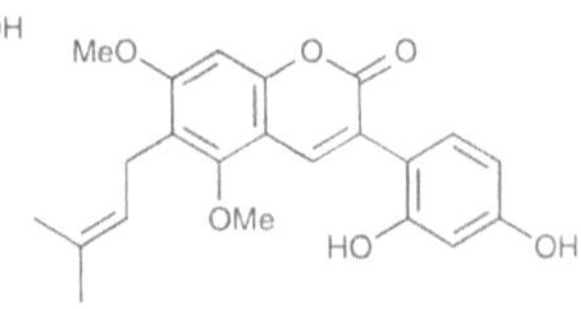

glycyrin (248)

Fig. 18

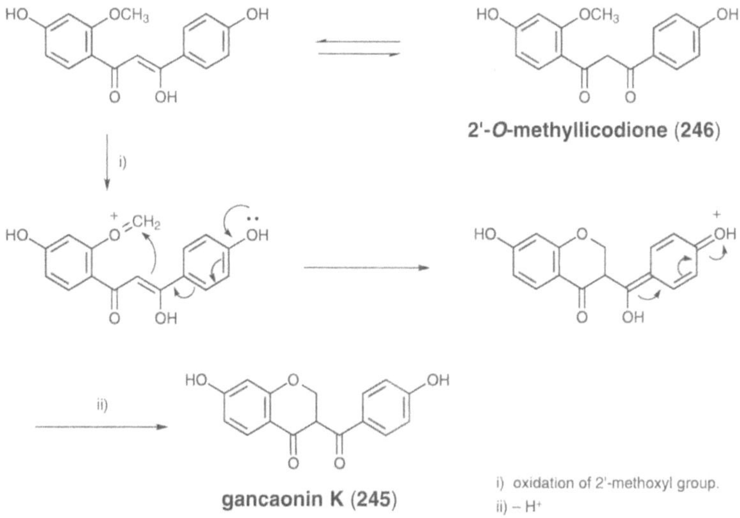

Scheme 1

material used in preparation of traditional Sino-Japanese medicine (*29*, *317*, *318*, *325–327*). Generally the traditional medicines consist of mixtures of crude drugs whose preparation is troublesome because it requires extraction with boiling water for lengthy periods. Until relatively recently the consumption of licorice was small enough to permit such methods. However since then the consumption of drugs has increased rapidly with a consequent increase in the use of licorice. While modern pharmaceutical industry supplies the traditional Chinese medicines (Kanpo-yaku in Japanese) in the form of powder prepared by lyophilization, it is not easy nowadays to obtain Northeast licorice. For this reason Northwest licorice is now used instead of Northeast licorice. The two licorices differ in their taste and morphology. By analysis of the sapogenins with gas-liquid chromatography, the ratio of uralenic acid (**22**, 18α-glycyrrhetic acid) to 18β-glycyrrhetic acid (**1**) differs in the two crude drugs (*328*). Attempts to identify the species represented by Northwest licorice have not yet been successful (*29*, *228*, *326*, *327*, *329–332*).

The main phenolic constituents of the two licorices are almost the same, but some minor compounds differ as is demonstrated by the following studies. About forty flavonoids have been isolated from Northwest licorice purchased in Japanese markets. Among these, sixteen phenolic constituents were new. The main phenolic compounds were liquiritin apioside (**115**, 0.33–2.51%), liquiritin (**110**, 0.20–1.23%) and licurazid (**120**, 0.1–0.65%) (*326*). A prenylflavonol, licoflavonol (**208**), and a prenylated 3-arylcoumarin, glycyrin (**248**), were isolated along with a prenylisoflavone, licoisoflavone A (**249**), by SHIBATA *et al.* (*283*, *333*, *334*). Three isoflavones with a prenyl group, glycyrrhisoflavone (**250**), glisoflavone (**251**) and glicoricone (**252**), a pyranoisoflavanone glycyrrhisoflavanone (**253**) and an isoflavanone with two prenyl groups (**254**) were isolated together with two 3-arylcoumarins containing an isoprenoid substituent, licoarylcoumarin (**255**) and licopyranocoumarin (**256**), and a furanone derivative licofuranone (**257**) from Northwest licorice by OKUDA *et al.* (*335–338*). We isolated a prenylated 2-arylbenzofuran gancaonin I (**177**, Fig. 13), a pyranocoumestan gancaonin F (**258**), two prenylated isoflavones gancaonins G (**259**) and H (**260**) and a prenylated 3-arylcoumarin gancaonin W (**261**) together with ten known flavonoids from Northwest licorice obtained from a Japanese market (*277*, *339*).

Among the flavonoids, compounds **250**–**253**, **257** and **258** have been isolated only from Northwest licorice and not from any botanically identified *Glycyrrhiza* species. Compounds **161**, **162**, **171**, **248**, **255** and **256** have been isolated from both Northeast and Northwest licorices, while compounds **177**, **208**, **209** and **259** were isolated from *G. aspera*, Northeast licorice (*G. uralensis*) and Northwest licorice. Gancaonin W (**261**) and kumatakenin (**262**) have been isolated from both *G. aspera* and Northwest licorice. Compound (**262**) is used as a criterion for identification of Northwest licorice (*318*, *331*). Judging from the minor phenolic components,* Northwest licorice may be a hybrid of *G. uralensis* and some other *Glycyrrhiza* species, such as *G. aspera*. There is also a possibility that Northwest licorice is a variety of *G. uralensis* or that the morphology of *G. uralensis* changes with location or growth conditions.

3.1.8. Tiexin Gancao, a Folk Medicine of Huizu

Tiexin gancao (hard core licorice) is one of the folk medicines in the Ningxia Huizu Autonomous Region of northwestern China. The crude

* Licoricidin (**164**) is one of the main flavonoids of *G. uralensis*, *G. aspera* and Northwest licorice, but could not be isolated from *G. uralensis* in the Xinjiang licorice of subsection 3.1.4. The constituents of the licorice are shown in Table 13 (unpublished data, as *G. uralensis* [Xinjiang]). Among them, 7-*O*-methylluteone (**263**) was isolated as a new natural product.

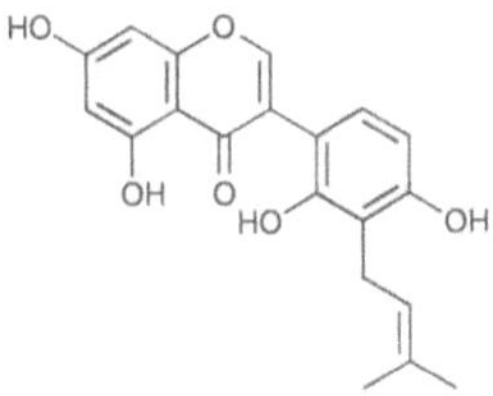

licoisoflavone A (249)

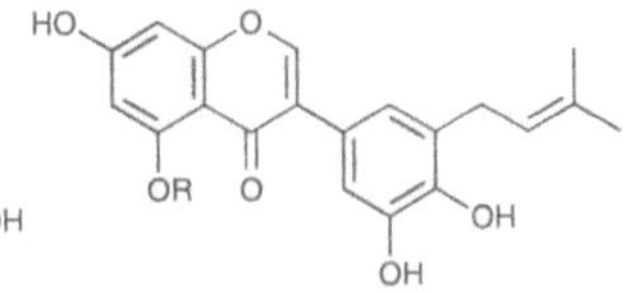

glycyrrhisoflavone (250):
R = H
glisoflavone (251):
R = Me

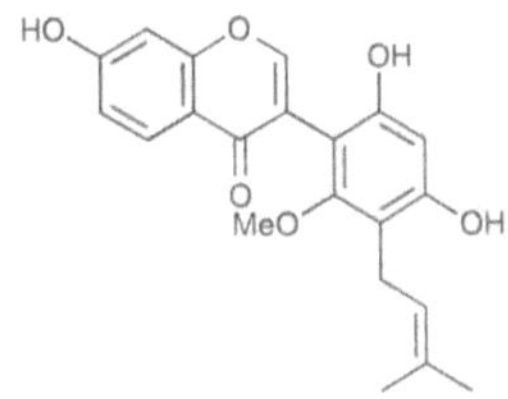

glicoricone (252)

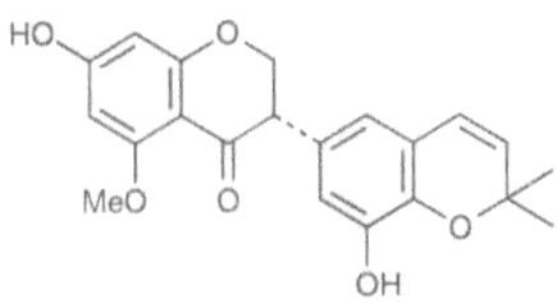

glycyrrhisoflavanone (253)

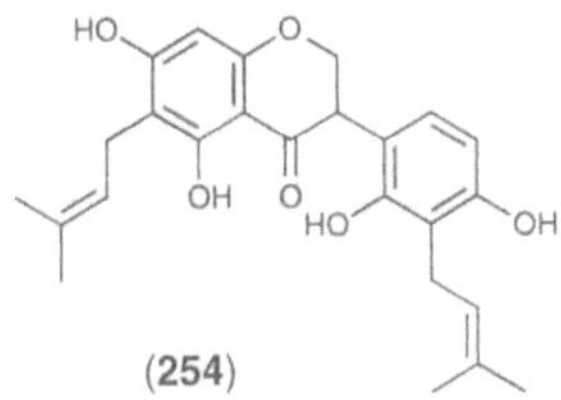

(254)

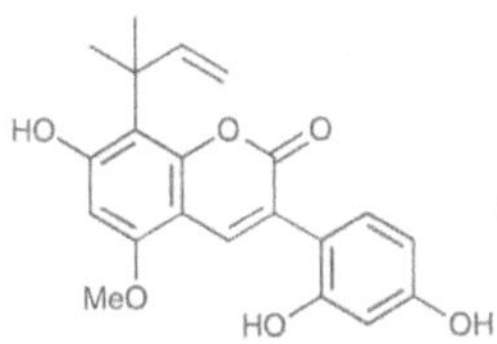

licoarylcoumarin (255)

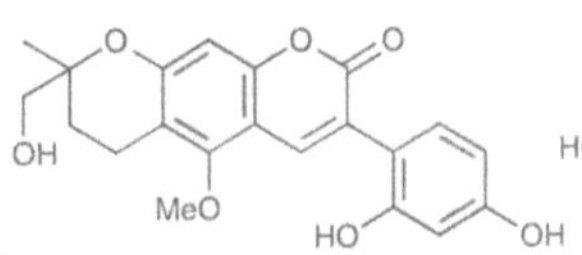

licopyranocoumarin (256)

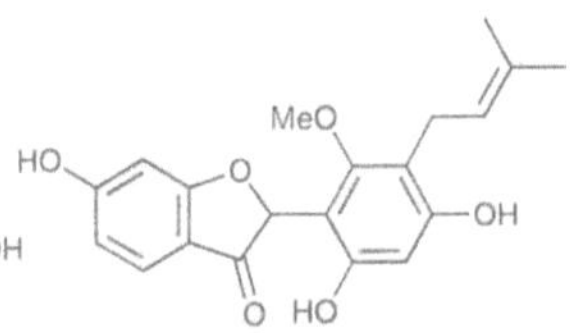

licofuranone (257)

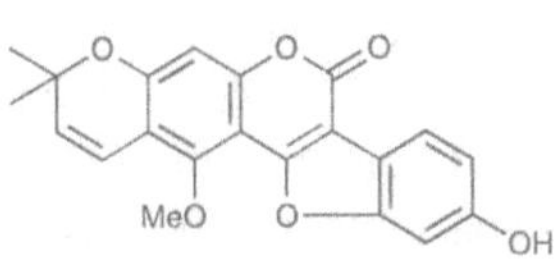

gacaonin F (258)

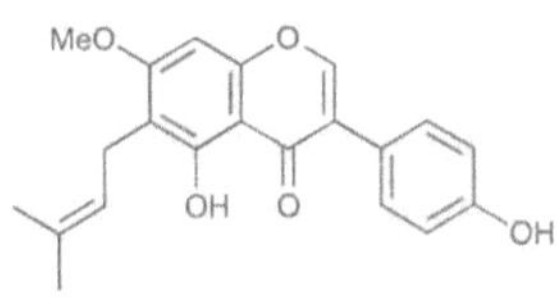

gancaonin G (259)

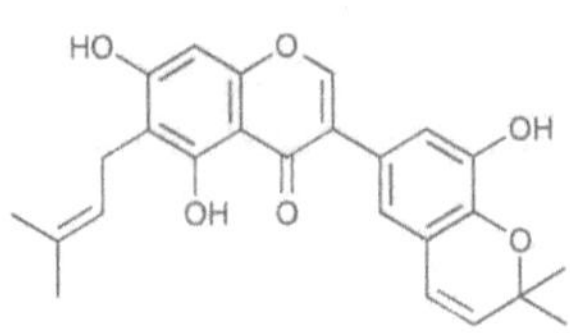

gancaonin H (260)

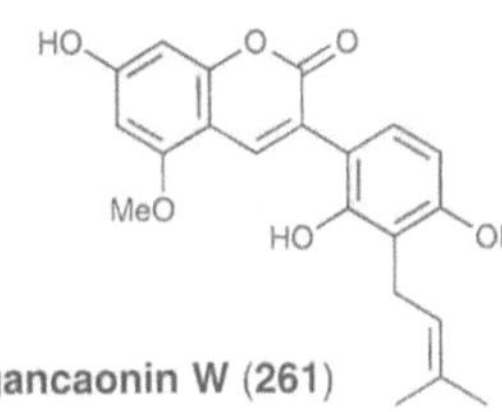

gancaonin W (261)

Fig. 19

drug is seldom found in *Glycyrrhiza* plants growing in that area and has been considered as being derived from diseased licorice. When hard cores are encountered in the manufacturing process which involves cutting both ends of the roots with knife, the cores are separated and called Tiexin Gancao. We isolated three new isoflavans, gancaonins X (**264**), Y (**265**) and Z (**266**), and fourteen known flavonoids from this drug (*340*). The known compounds (Table 13) had already been isolated previously from licorice species growing in northeastern China (*G. eurycarpa*, *G. inflata*, *G. glabra*, *G. aspera* and *G. uralensis*).

3.2. Phenolic Compounds from Aerial Parts

There are a number of reports dealing with the phenolic constituents of the aerial parts of *Glycyrrhiza* species. About thirty new prenylated phenols have been isolated; these compounds are not generally found in the underground parts.

3.2.1. Glycyrrhiza glabra

About thirty phenolic compounds have been isolated or detected by HPLC analysis. Two prenylflavanones, glabranin (**267**) and licoflavanone (**268**), have been isolated by KATTAEV and NIKONOV (*226*), and TABATA *et al.* (*341*). LITVINENKO and NADEZHINA isolated the *cis*-3-hydroxy-flavanone folerogenin (**269**) from the species (*342*, *343*). Three new glucosides, glyphoside (**270**), kaempferol-3-*O*-diglucoside (**271**) and astragalin (kaempferol-3-*O*-glucoside) monoacetate (**272**), have been isolated, but the structure of the glycosyl moiety of **271** and the position of the acetate of **272** have not been elucidated (*342*, *344*).

INGHAM reported an isoflavan phytoalexin in the leaves of this species (*345*). The phytoalexin isomucronulatol (**273**) is produced on inoculation with *Helminthosporium carbonum*.

3.2.2. Glycyrrhiza uralensis

Forty phenolic compounds have been isolated from the aerial parts. We isolated twenty-one phenolic compounds from the aerial parts collected in Heilogjang Province in northeastern China (*346–348*). Among these, gancaonin A (**274**) was isolated for the first time as a natural product and fifteen compounds, gancaonins B–E and L–V (**275–289**) were isolated as new compounds. Among these, gancaonins A (**274**), B (**275**), L (**279**), M (**280**) and N (**281**) are prenylated isoflavones,

References, pp. 110–140

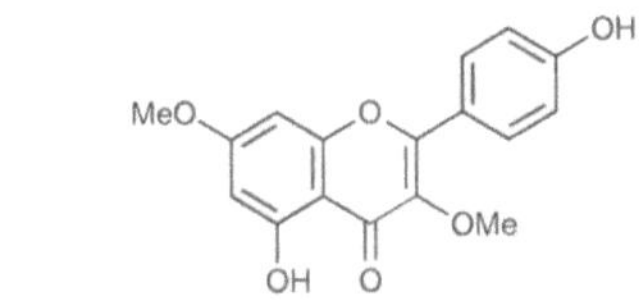

kumatakenin (262)

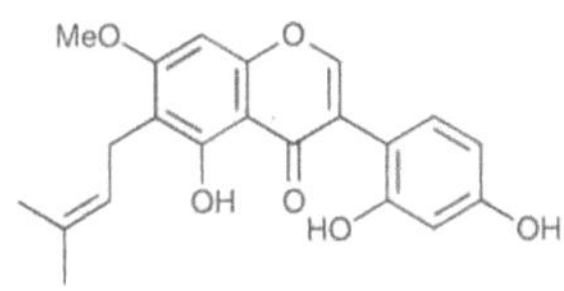

7-*O*-methylluteone (263)

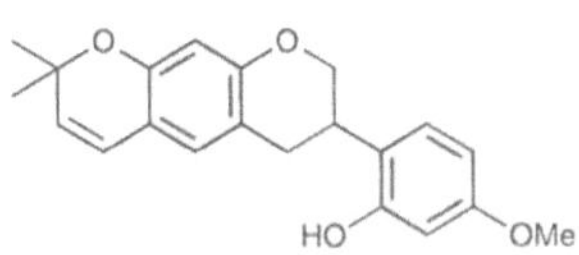

gancaonin X (264)

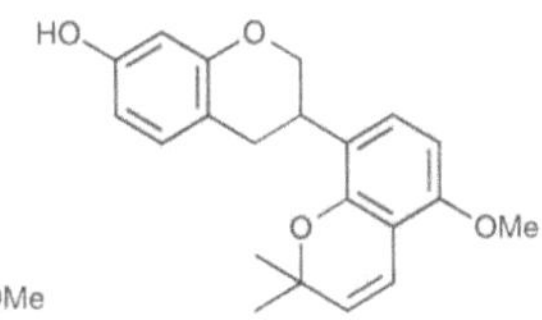

gancaonin Y (265)

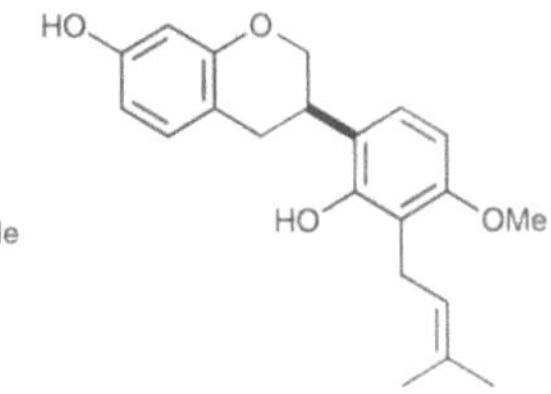

gancaonin Z (266)

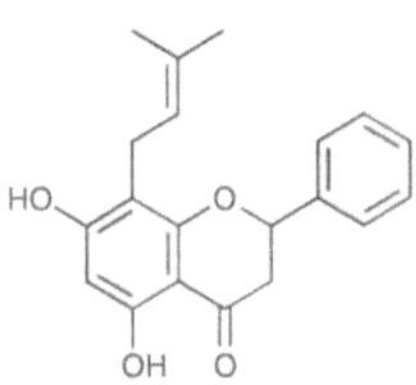

glabranin (267)

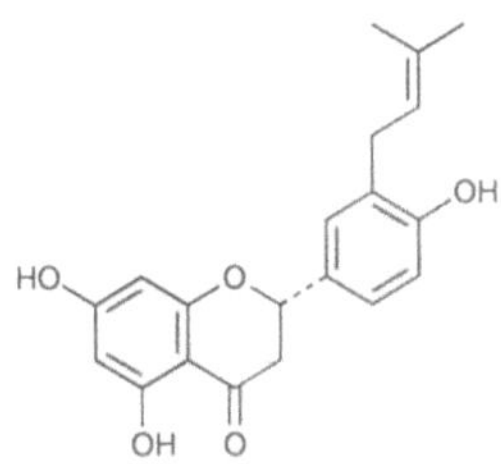

licoflavanone (268)

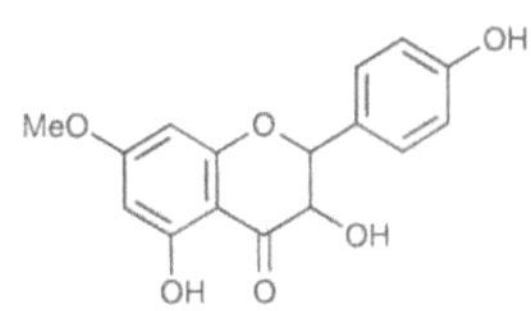

folerogenin (269)

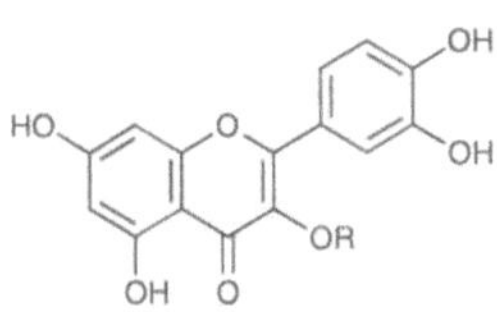

glyphoside (270):
R = ($CH_3CO \rightarrow 2$)Glu-

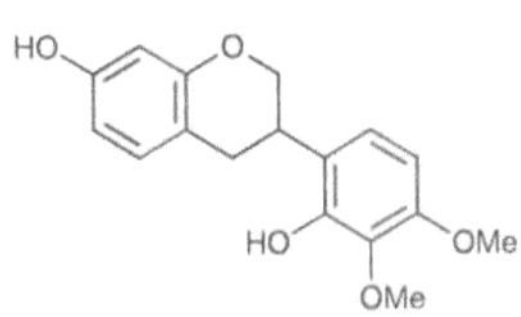

isomucronulatol (273)

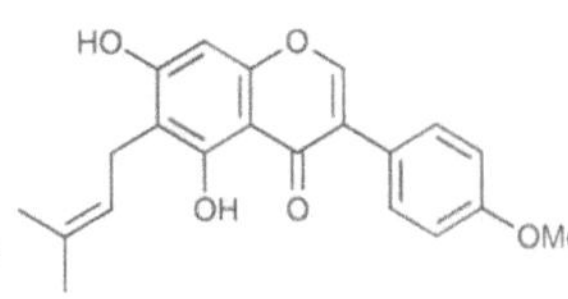

gancaonin A (274)

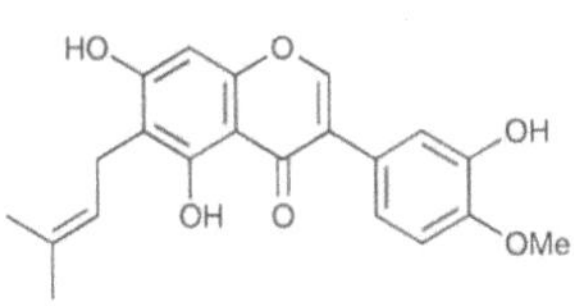

gancaonin B (275)

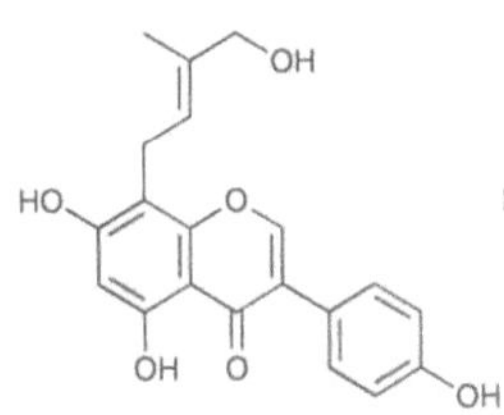

gancaonin C (276)

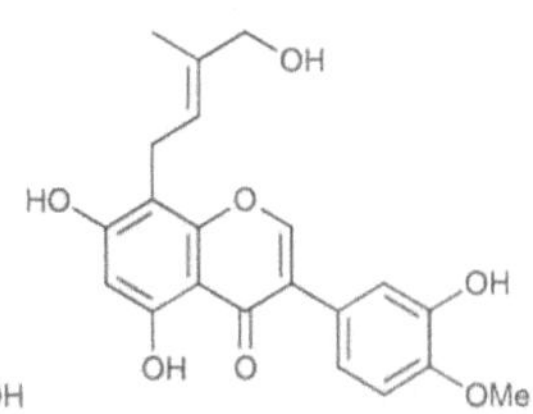

gancaonin D (277)

Fig. 20

gancaonins O (**282**) and Q (**284**) are prenylated flavones, and gancaonin P (**283**) is a prenylated flavonol.

During structure elucidation of these flavonoids, we found characteristic fragmentation patterns involving cleavages of prenyl groups as shown in Scheme 2 (*347*). The EI-mass spectra of 6-prenylated flavonoids give rise to a strong peak at M – 43 (C_3H_7) together with a strong peak at M – 55 (C_4H_7) and weak peaks at M – 15 (CH_3) and M – 68 (C_5H_8) but the characteristic peak at M – 43 is not observed in the spectra of 8-prenylflavonoids. These compounds give rise to a strong peak at M – 15 (CH_3) along with medium-sized peaks at M – 55 (C_4H_7) and M – 68 (C_5H_8). This difference in fragmentation patterns is useful for discriminating between the two classes of compounds and is

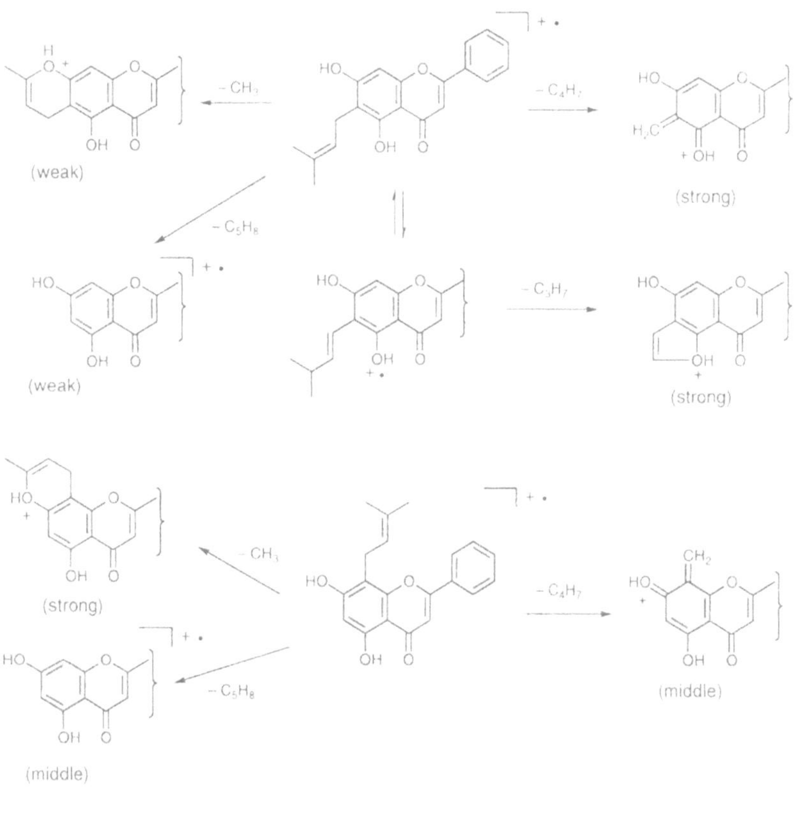

Scheme 2

characteristic of monoprenylated flavonoids. On the other hand, 6- and 8-prenylated flavanones do not show such characteristic fragmentation patterns because they exist as equilibrium mixtures of chalcones and flavanones under EI conditions. However, one can discriminate between them by means of their FAB-mass spectra (*349*). Details of the mass spectrometry of prenylated flavonoids have been reviewed (*350*).

Gancaonins C (**276**) and D (**277**) are isoflavones with an (*E*)-3-hydroxymethyl-2-butenyl group at C-8. Gancaonins R (**285**), S (**286**) and T (**287**) are prenylated dihydrostilbenes and gancaonins U (**288**) and V (**289**) are prenylated dihydrophenanthrenes. The ^{13}C NMR spectra were a useful tool in the structure determination of these prenylated compounds. The substituent patterns of the hydroxyl groups easily follow from the chemical shifts of the aromatic carbons (*351*). The nature of the substituents *ortho* to a prenyl group can be determined by the chemical shift of the benzylic methylene carbon in the prenyl group. These shifts are useful for structure determination of prenylphenols especially for complex compounds. Details have already been reported in a review article (*352*), but the regularities are described briefly in Section 4.

Five prenylflavonols have been isolated from the leaves of the plant by JIA *et al.* (*353–356*). Among these, three are flavonols of common type, *i.e.*, uralenol (**290**), uralenol 3-*O*-methyl ether (**291**), and gancaonin P 3′-*O*-methyl ether (**283a**), but in two flavonols, *i.e.*, neouralenol (**292**) and uralene (**293**), the substituent patterns of the hydroxyl group and/or a prenyl group are unusual. JIA *et al.* have also reported a new phenol glycoside, 1-*O*-protocatecuyl-β-D-xylose (uralenneoside), together with six known glycosides from the leaves (*357*).

3.2.3. Glycyrrhiza acanthocarpa

GHISALBERTI *et al.* have reported the isolation of two dihydrostilbenes (**294** and **295**) having a prenyl group and two prenylphenols (**296** and **297**) from the leaves and terminal branches of *G. acanthocarpa* (Lindl) J.B. Black collected in Western Australia (*358*).

3.2.4. Glycyrrhiza eurycarpa

We examined the aerial parts of *G. eurycarpa* collected in Gansu Province of northwestern China, and isolated a new isoprenoid-substituted flavanone, kanzonol S (**298**), together with thirteen known flavonoids (*359*). Kanzonol S (**298**) is the first example of a flavonoid with an (*E*)-3-acetoxymethyl-2-butenyl group.

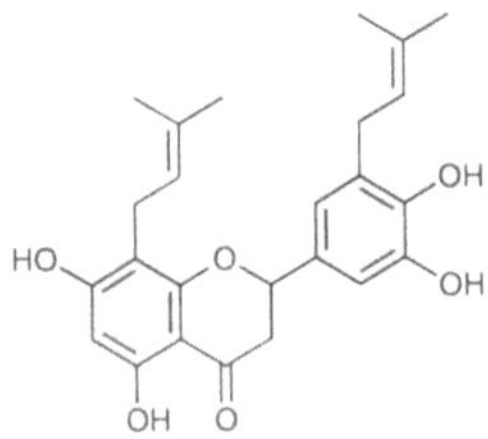

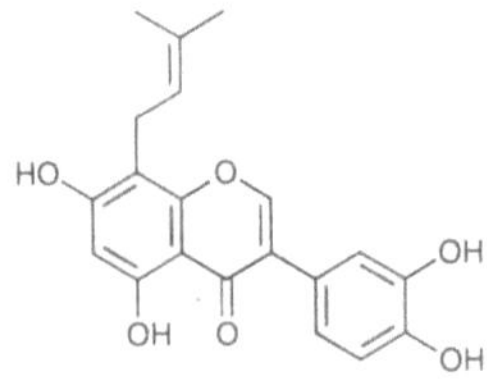

gancaonin L (279)

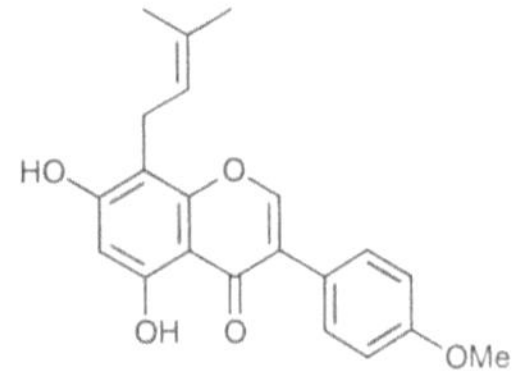

gancaonin M (280)

gancaonin E (278)

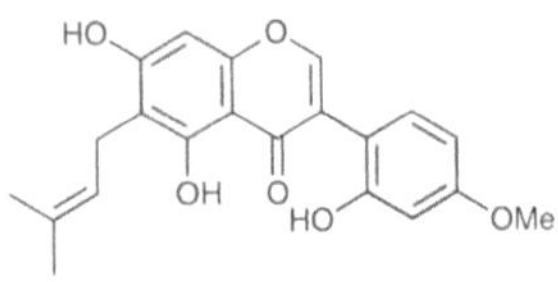

gancaonin N (281)

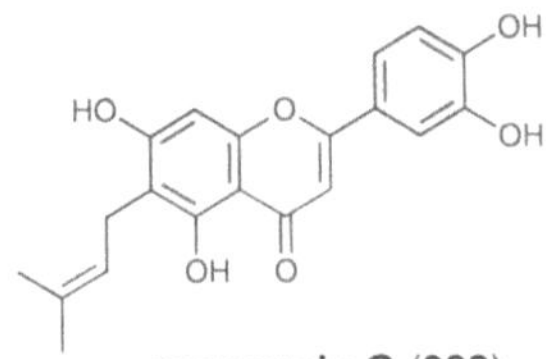

gancaonin O (282)

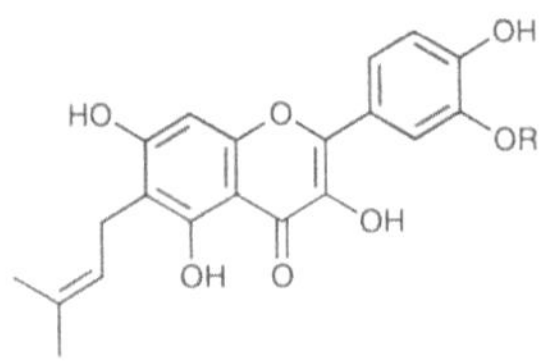

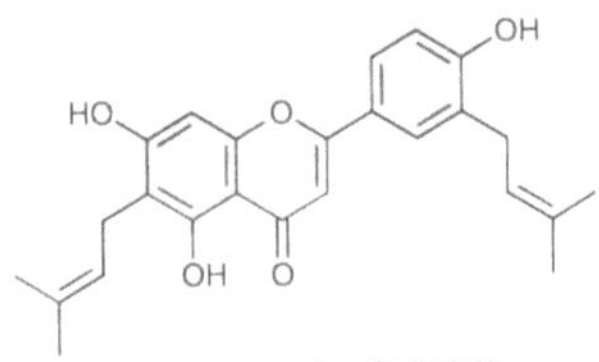

gancaonin Q (284)

gancaonin P (283):
R = H
gancaonin P 3'-*O*-methyl ether (283a):
R = Me

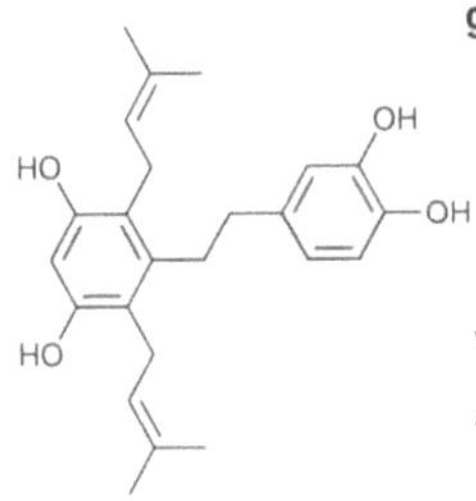

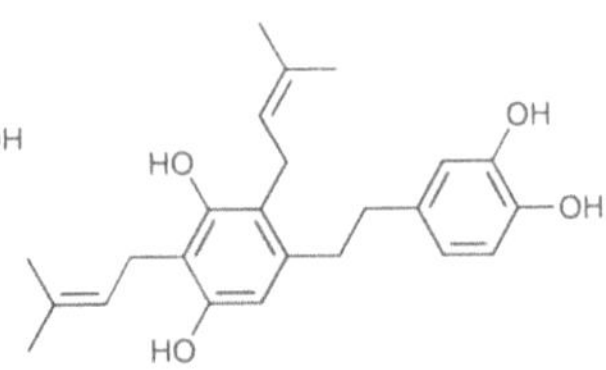

gancaonin S (286)

gancaonin R (285)

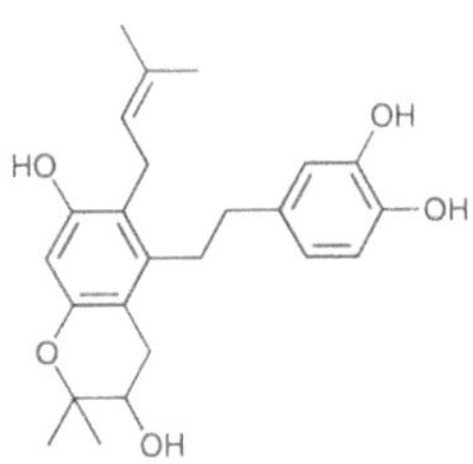

gancaonin T (287)

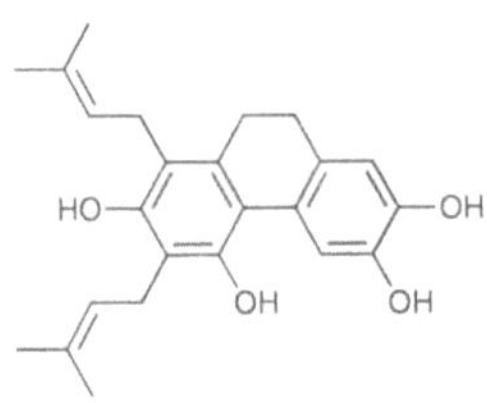

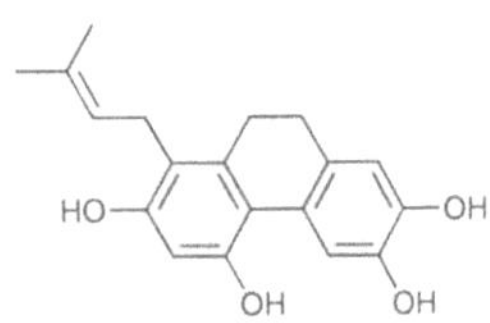

gancaonin V (289)

gancaonin U (288)

Fig. 21

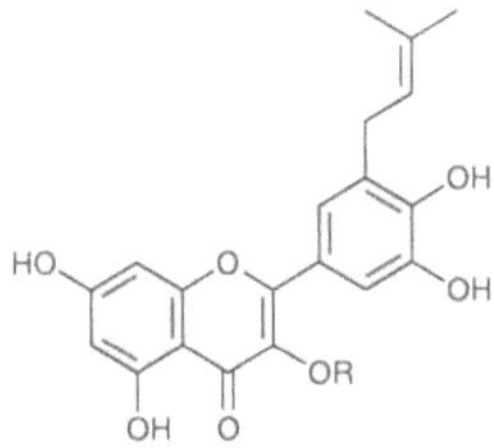

uralenol (290): R = H
uralenol 3-*O*-methyl ether (291): R = Me

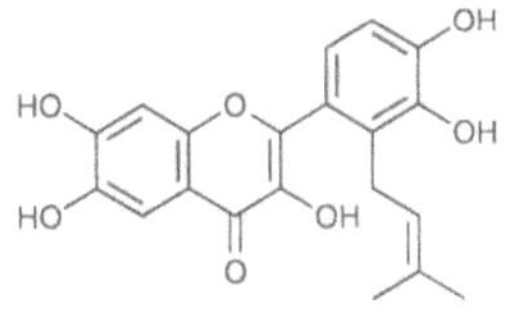

neouralenol (292)

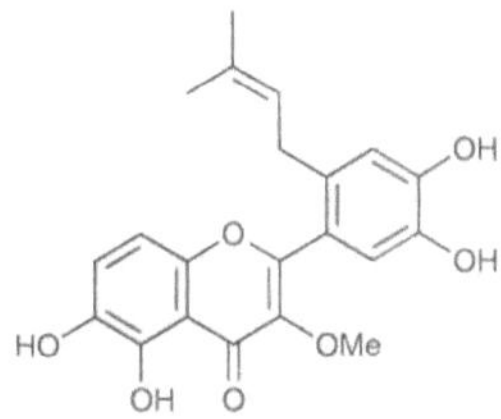

uralene (293)

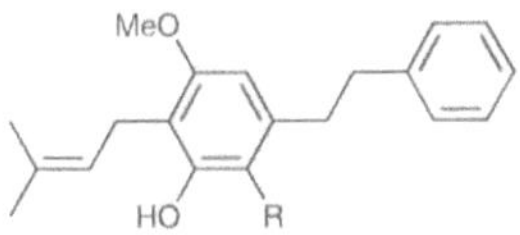

(**294**): R = H
(**295**): R = COOH

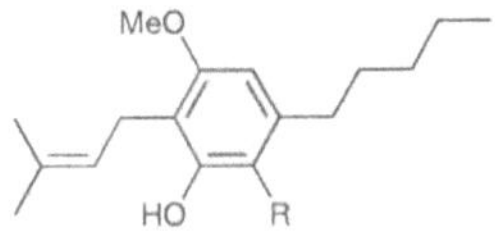

(**296**): R = H
(**297**): R = COOH

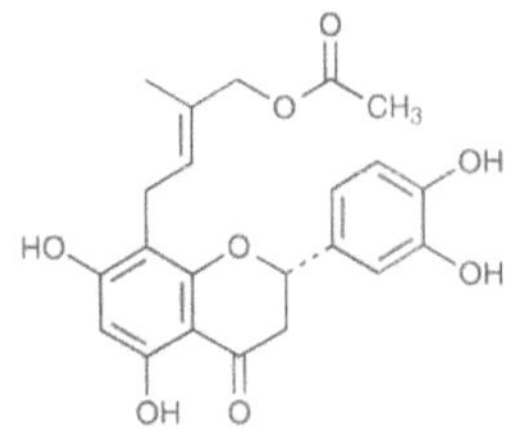

kanzonol S (298)

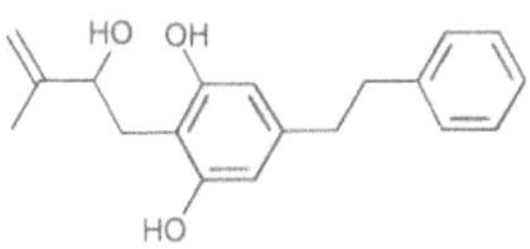

glepidotin C (299)

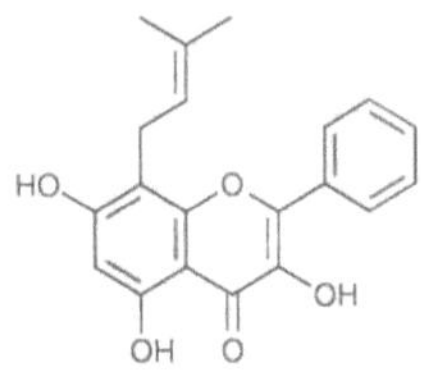

(**300**)

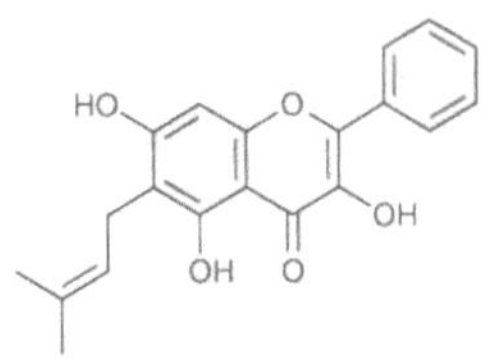

glepidotin A (301)

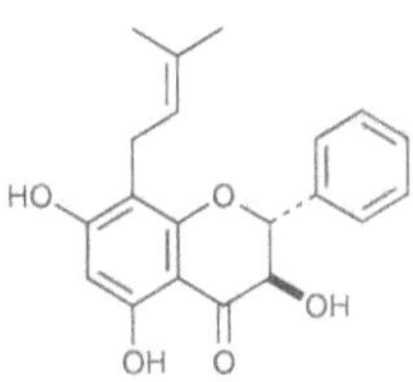

(**302**)

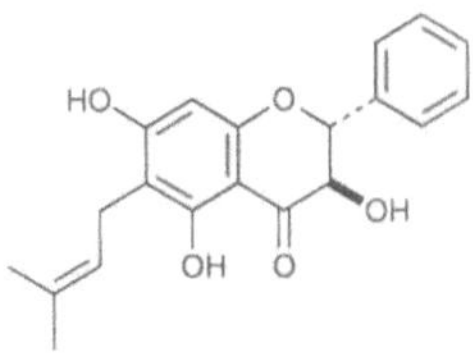

glepidotin B (303)

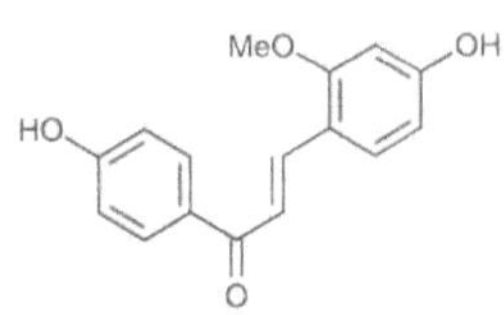

echinatin (304)

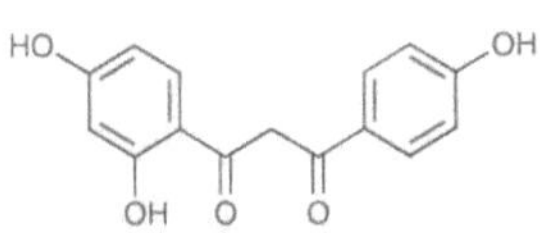

licodione (305)

Fig. 22

3.3. Phenolic Compounds from Whole Plants

An optically active dihydrostilbene having a 2-hydroxy-3-methyl-3-butenyl group, glepidotin C (**299**), has been isolated together with two prenylated flavonoids (glepidotins A and B) from the whole plant of *G. lepidota* (American or wild licorice) by MITSCHER *et al.* (*360*, *361*). The structure of glepidotin A was postulated to be that shown in formula **300** (5,7-dihydroxy-8-prenylflavonol) because the naturally occurring compound was different from a presumed 5,7-dihydroxy-6-prenylflavonol (**301**) whose synthesis had been reported earlier by JAIN and ZUTSHI (*362*). However, the synthetic compound was assigned its structure on insecure evidence. Glepidotin B was assigned formula **302** by correlating it with glepidotin A since glepidotin B triacetate was converted to glepidotin A diacetate with chromous chloride in acetone. However, as a result of our systematic study of the chemical shifts of hydrogen-bonded hydroxyl protons of flavonoids, the chemical shifts of glepidotins A and B did not fit the postulated structures. The details will be described in Chapter 4. We have synthesized the authentic prenylflavonols **300** and **301** whose structures were confirmed by ^{13}C NMR spectrometry (*276*). Glepidotin A was thus identified as the 6-prenylflavonol **301**; therefore the structure of glepidotin B was revised to **303**.

3.4. Phenolic Compounds from Cell Cultures

FURUYA and co-workers have investigated cell cultures of *G. echinata* (*363–367*). A new reversely constructed chalcone echinatin (**304**) was isolated together with two dibenzoylmethanes, licodione (**305**) and 5′-prenyllicodione (**243**), and a prenylflavone, licoflavone A (**241**), from the same cell lines. These flavonoids have also been isolated from some intact *Glycyrrhiza* plants as mentioned earlier (Table 13). Studies of the biosynthesis of flavonoids in *G. echinata* cell cultures have been reported by the groups of FURUYA and/or AYABE (*297*, *314*, *368–376*), with the biosynthesis postulated to proceed as shown in Scheme 3 (*86*). The key intermediate, 2′-*O*-methyllicodione (**246**), in the biosynthesis of echinatin (**304**) has not been isolated from the cell lines, but was synthesized using licodione *O*-methyltransferase isolated from cell cultures of *G. echinata* (*314*).

Cell cultures of *G. uralensis* and *G. glabra* also produce echinatin (**304**, Table 13). Licodione (**305**) has also been isolated from cultured alfalfa cells stimulated by a naphthoquinone or yeast extract (*377*, *378*).

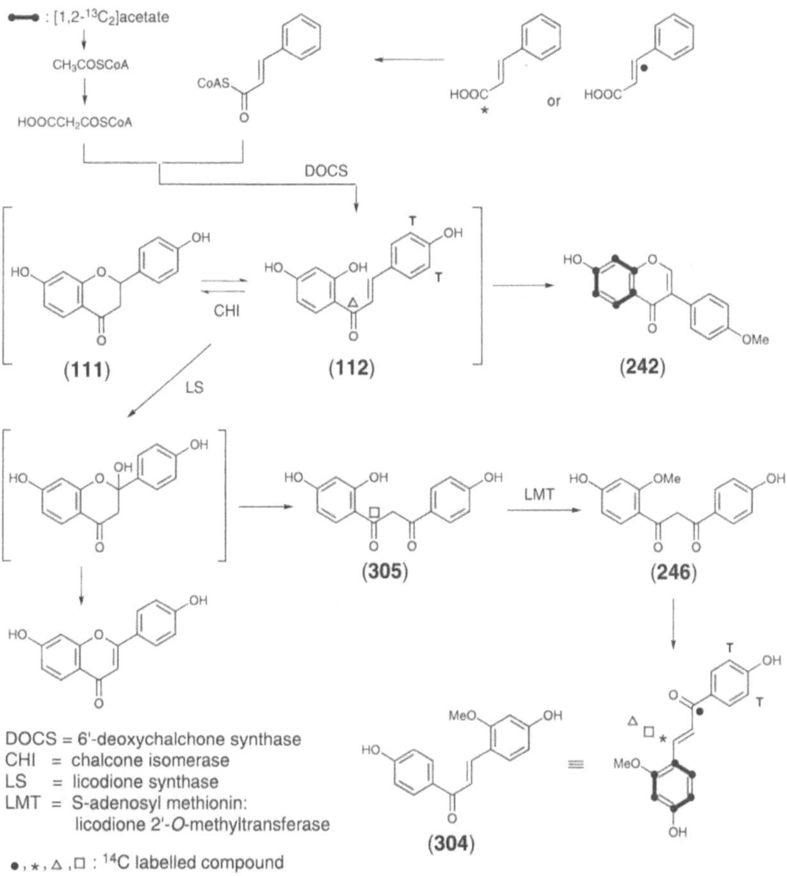

Scheme 3

4. Recent Methods of Structure Determination of Prenylated Phenols

4.1. Variations in the Chemical Shift of the Methylene Carbon of a Prenyl Group

The benzylic methylene carbon signal of the prenyl or (*E*)-3,7-dimethyl-2,6-octadienyl (geranyl) group (C1) of phenols appears between δ 20–30. In the course of our work on prenylated phenols, we noticed that the chemical shift of the C1 signal depends on the nature of the substituents located at the adjacent positions (*382*). To generalize the observation, we examined the ^{13}C NMR data of prenylated and/or geranylated phenols reported in the literature in addition to the phenols isolated or synthesized by our group. On this basis, the prenyl (geranyl) groups could be classified into six types (types 1–6) depending on the substituents located at the adjacent positions as follows (see Fig. 23): The prenyl group of type 1 lies between two oxygen functions. The prenyl group of type 2 is attached to C-3 of the flavone. Such 3-prenylflavones have been obtained only from Moraceous plants (*383–385*). In type 3 phenols an oxygen function is located at one of the positions *ortho* to the prenyl group and the other *ortho*-position is substituted with an alkyl or alkenyl group. In compounds of type 4 one of the positions *ortho* to the prenyl group is unsubstituted and an oxygen function is located at the other *ortho*-position. In type 5 compounds both positions *ortho* to the prenyl group are replaced by alkyl and/or alkenyl groups. Type 6 prenyl groups are found in xanthone derivatives in which the prenyl group is attached to C-8 of 7-oxygenated xanthones, *i.e.*, is located between carbonyl and hydroxyl groups.

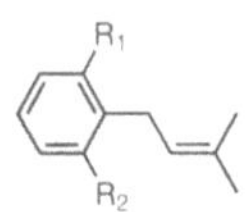

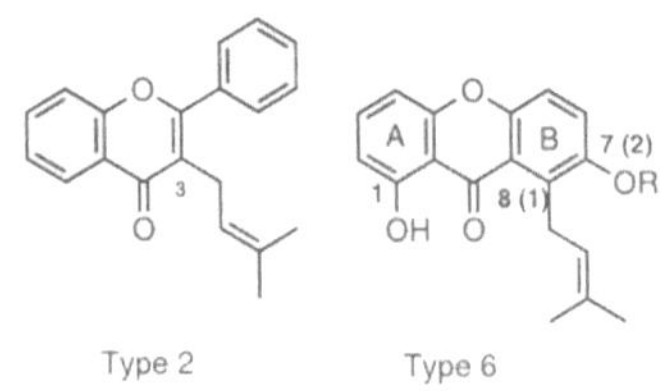

	R_1	R_2
Type 1	OH, *O*-alkyl, *O*-alkenyk	OH, *O*-alkyl, *O*-alkenyl
Type 3	OH, *O*-alkyl	alkyl, alkenyl
Type 4	OH, *O*-alkyl	H
Type 5	alkyl, alkenyl	alkyl, alkenyl

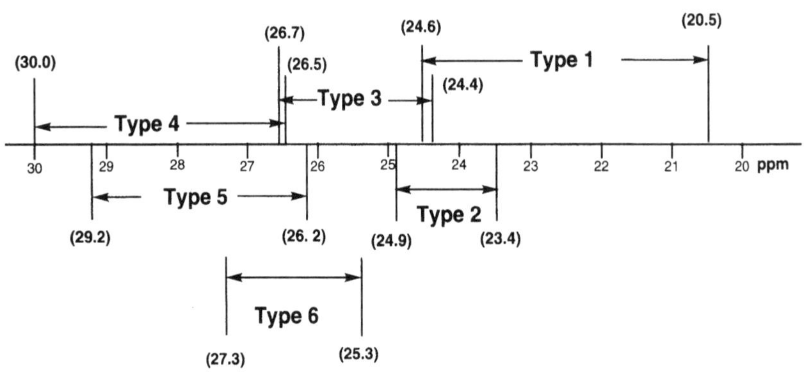

Fig. 23

The chemical shifts of the C1 signals of these six types of prenyl groups are observed in a restricted range specific to each type, except for compounds of types 4 and 5 which overlap as shown in Fig. 23. To differentiate between compounds of type 4 and type 5, it is necessary to determine whether there is spin-spin coupling between the methylene carbon and an aromatic proton ($^3J = 4\,\mathrm{Hz}$). Prenyl groups of type 5 are rarely found in natural compounds. Generally, such phenolic compounds have two isoprenoid substituents in one ring, and these groups are adjacent to each other as in broussoflavonol D (**306**) and kazinol P (**307**) (Fig. 24) (*382*, *386*).

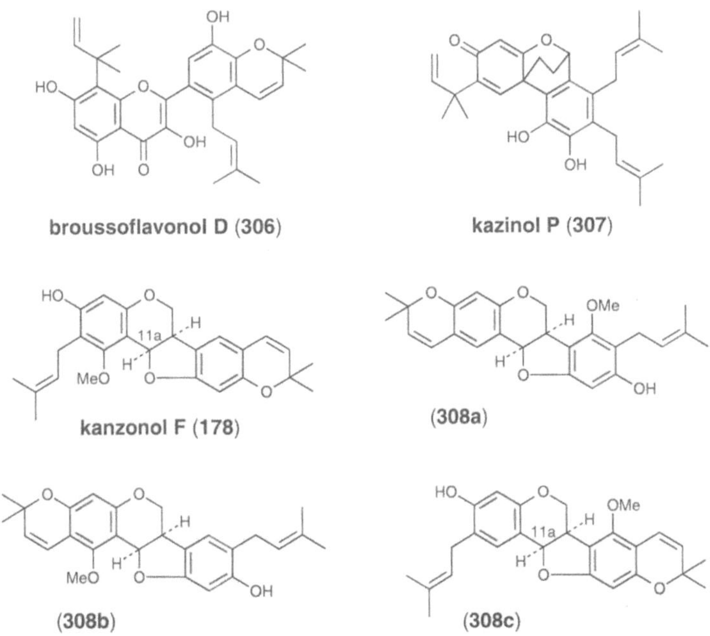

Fig. 24

In the above classification of chemical shifts, the boundaries are somewhat ambiguous (Fig. 23). Nevertheless, the boundaries are sharp when the chemical shifts are compared in the same solvent (*352*, *387*). Furthermore, among compounds with the same carbon skeleton, the methylene signals of each group occur within narrower regions (Table 3). The C1 chemical shifts of such flavonoids, coumarins and xanthones have been listed in a review article (*352*).

Table 3. *The Chemical Shifts of Methylene Carbons of Prenyl (Geranyl) Groups of Phenolic Compounds (δ in ppm)*

	Flavones, flavonols, isoflavones, chromones and aurones		
Type 1	21.6–22.6 (A ring)	21.2–22.7 (A ring)	20.8–21.9 (A ring)
	23.2–23.8 (B ring)		22.2–22.6 (B ring)
Type 2	24.4–24.7	24.1–24.4	23.4–24.1
Type 3	25.8–27.3	25.8–26.2	25.2
Type 4	28.6–29.5	28.9–29.2	27.4–28.2
Type 5		28.0–29.2	27.3–28.4
Solvent	Acetone-d_6	$CDCl_3$	DMSO-d_6
	Flavanones, 3-hydroxyflavanones (dihydroflavonols) and isoflavanones		
Type 1	21.4–22.8 (A ring)	20.8–22.8 (A ring)	20.5–22.3 (A ring)
	23.3–24.6 (B ring)	—	23.0–23.3 (B ring)
Type 3	24.9–25.6	—	—
Type 4	27.6–29.1	27.6–29.8	26.9–29.1
Solvent	Acetone-d_6	$CDCl_3$	DMSO-d_6
	Flavans, catechins, isoflavans and pterocarpans		
Type 1	22.6–22.8 (catechins)	22.0–22.7 (catechins)	—
	23.1–23.6	22.5–23.2	—
Type 3	—	25.1–25.6	—
Type 4	28.7–28.8	28.6–29.3	—
Solvent	Acetone-d_6	$CDCl_3$	DMSO-d_6
	2-Arylbenzofurans, coumaronochromones, 3-arylcoumarins, coumestans and coumarins		
Type 1	22.0–24.0	21.6–23.5	21.0–22.3
Type 3	—	25.8–26.7	—
Solvent	Acetone-d_6	$CDCl_3$	DMSO-d_6
	Xanthones		
Type 1	21.6–22.8 (A ring)	21.2–23.3 (A ring)	20.7–21.2 (A ring)
	23.0–23.1 (B ring)	—	—
Type 4	28.8–29.0	27.4–28.1	28.3–28.6
Type 6	26.0–27.3	25.6–26.8	25.3–25.8
Solvent	Acetone-d_6	$CDCl_3$	DMSO-d_6
	Chalcones, dihydrochalcones, acetophenons, benzophenones, benzoic acids, coumaric acids, 1,3-diphenylpropanes, stilbenes, dihydrostilbenes and dibenzoylmethanes		
Type 1	21.5–23.2	21.4–23.4	—
Type 3	25.5–26.2	24.4–26.1	—

Table 3 (*continued*)

Type 4	28.2–29.2	28.4–30.0	—
Type 5	—	26.2–27.8	—
Solvent	Acetone-d_6	$CDCl_3$	DMSO-d_6
	Phenols and dihyrophenancerene		
Type 1	23.7	23.6–24.6	—
Type 2	25.3–25.9	25.5–25.9	—
Type 4	28.3	26.5–29.8	—
Solvent	Acetone-d_6	$CDCl_3$	DMSO-d_6

This method is useful for characterization of complex phenolic compounds. For example, the ^{1}H NMR spectrum of kanzonol F (**178**) allows for postulation of twenty possible structures. However, the chemical shift of the methoxyl carbon (δ 63.1 in acetone-d_6) reduces the number of the possible structures to four (**178** and **308a–c**, Fig. 24) since the signal of *ortho*-disubstituted methoxyl carbon appears near δ 60 whereas that of an *ortho* unsubstituted methoxyl carbon is near δ 55 (*388*, *389*). Formulas **308b** and **308c** are excluded by the chemical shift of the C1 signal of the prenyl group at δ 23.4 in acetone-d_6 whereas if the structure of kanzonol F were **308b** or **308c**, the C1 frequency of type 4 should have occurred between δ 26.5 and 30.0. Finally, structure **178** was differentiated from structure **308a** by observation of an NOE between the methoxyl and H-11a. Using the C1 chemical shift method structures of several naturally occurring compounds could be revised (*382*, *390–392*). The same method would probably be useful for structure determination of phenolic alkaloids containing prenyl residues (*393*).

4.2. Variations in the Chemical Shift of the 5-Hydroxyl Proton of 6- or 8-Prenylated Flavonoids

Many 5,7-dihydroxyflavonoids with prenyl or geranyl groups at the C-6 or C-8 position have been isolated from natural sources including licorice (*394–397*). Various methods exist for determining the position of substitution, such as (*a*) the Gibbs test for flavonoids having no hydroxyl group at C-2′ or C-3′, (*b*) the anomalous aluminum induced UV shift, (*c*) cyclization reactions (6-prenylflavonoids afford two products because of cyclization to 5-OH or 7-OH, 8-prenylflavonoids afford only one product by cyclization to 7-OH), (*d*) chemical shifts of the vinyl methyl signals of the prenyl group in the ^{1}H NMR spectra (this is only appropriate for flavanones), (*e*) chemical shifts of an aromatic proton on ring A and the methylene protons of the prenyl group (useful for flavones and isoflavones), (*f*) shifts of chromene protons after acetylation of a 5-hydroxyl group (for this opportunity for a cyclization reaction must exist), (*g*) chemical shifts of C-6 and C-8 carbons, *etc.* (*398–405*). These methods are not always applicable as shown in Table 4 and may therefore lead to errors. Some examples of such errors are described in the following paragraphs and in Section 3.

Table 4. *The Examples of Unsuitable Method for the Structure Determination of 6- or 8-Prenylated Flavonoids*

Methods of structure determination	Unsuitable flavonoid	Ref.
(*a*) Gibbs test (*398*)	Flavonol (coumarin)	(*276*)
(*b*) Aluminum-induced shift (*399*)	Flavone (except 3-prenylflavone), flavanone	(*285*)
(*c*) Cyclization reaction (*400*)	Flavanone (Wessely-Moser rearrangement)	(*413*)
(*d*) Chemical shift of vinyl methyl proton (*401*)	(only suitable for flavanone)	
(*e*) Chemical shift of C-6-H or C-8-H and methylene protons (*402*)	Flavanone, isoflavanone	(*262, 402*)
(*f*) Acetylation shift on chromeno protons (*403*)	2′-Hydroxyflavone	(*402*)
(*g*) Chemical shift of C-6 and C-8 (*404, 405*)	Flavanone, isoflavanone	(*402, 408*)

An unambiguous method for the structure determination of such prenylated flavonoids is the observation of spin-spin coupling between the 5-hydroxyl proton and C-6 ($^{3}J = 6\,Hz$) by a ^{13}C NMR measurement such as an undecoupled, HMBC spectrum (*406–408*). This method needs relatively large amounts of a sample, but the following ^{1}H NMR method requires only a small amount of the flavonoid (less than 200 μg for routine measurements).

4.2.1 Flavanones

In the course of our work on the flavonoids from licorice, we noticed that the chemical shift of the hydrogen-bonded hydroxyl proton on C-5 of flavonoids is shifted by prenylation at C-6 or C-8 (*347*). On examining the ^{1}H NMR data of fifty isoprenoid-substituted flavanones reported in the literature and the flavanones isolated or synthesized by our group (*409*) where the hydrogen-bonded hydroxyl proton (5-OH) appears between δ 12.0–12.6 (in acetone-d_6) and δ 11.9–12.4 (in $CDCl_3$) one finds that the 5-OH of 6-prenylated flavanones appears further downfield than that of 6-unsubstituted flavanones having the same B ring. In contrast, the 5-OH signal of an 8-prenylated flavanone is shifted slightly upfield compared with that of a flavanone having the same B ring and no side chain. The same regularities were observed in similar flavanones containing other isoprenoid groups such as geranyl, isoamyl, 3-hydroxy-3-methylbutyl and lavandulyl residues, *etc.* The downfield shift of 6-prenylated flavanones is about 0.3 ppm, and the upfield shift of 8-prenylated flavanone is about 0.05 ppm when the ^{1}H NMR spectra are measured in acetone-d_6; the downfield shift is 0.35 ppm when measured in $CDCl_3$ (*409*, *410*).

In examining the literature we found that the above generalization did not accord with the chemical shifts reported for six isoprenoid-substituted flavanones (*409*). Some of these may have been due to typing errors or a slip in recording of the spectrum. Recently two of these flavanones and some analogs were re-isolated or synthesized, with the

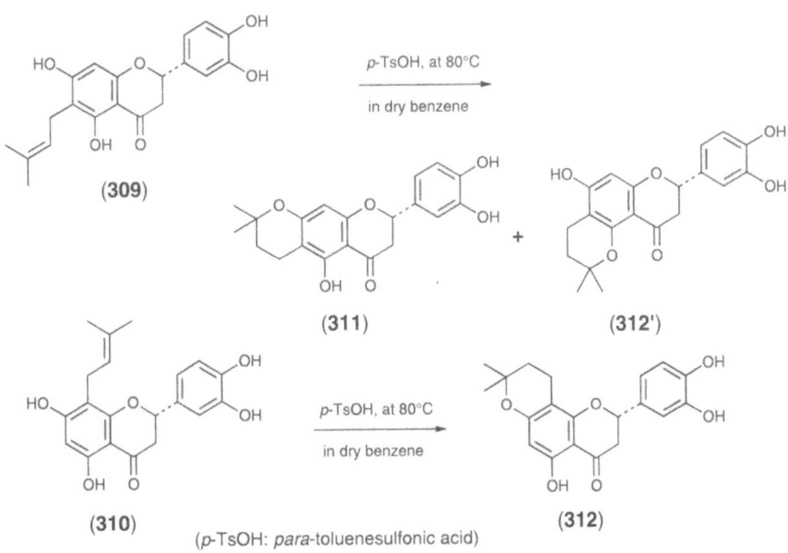

Scheme 4

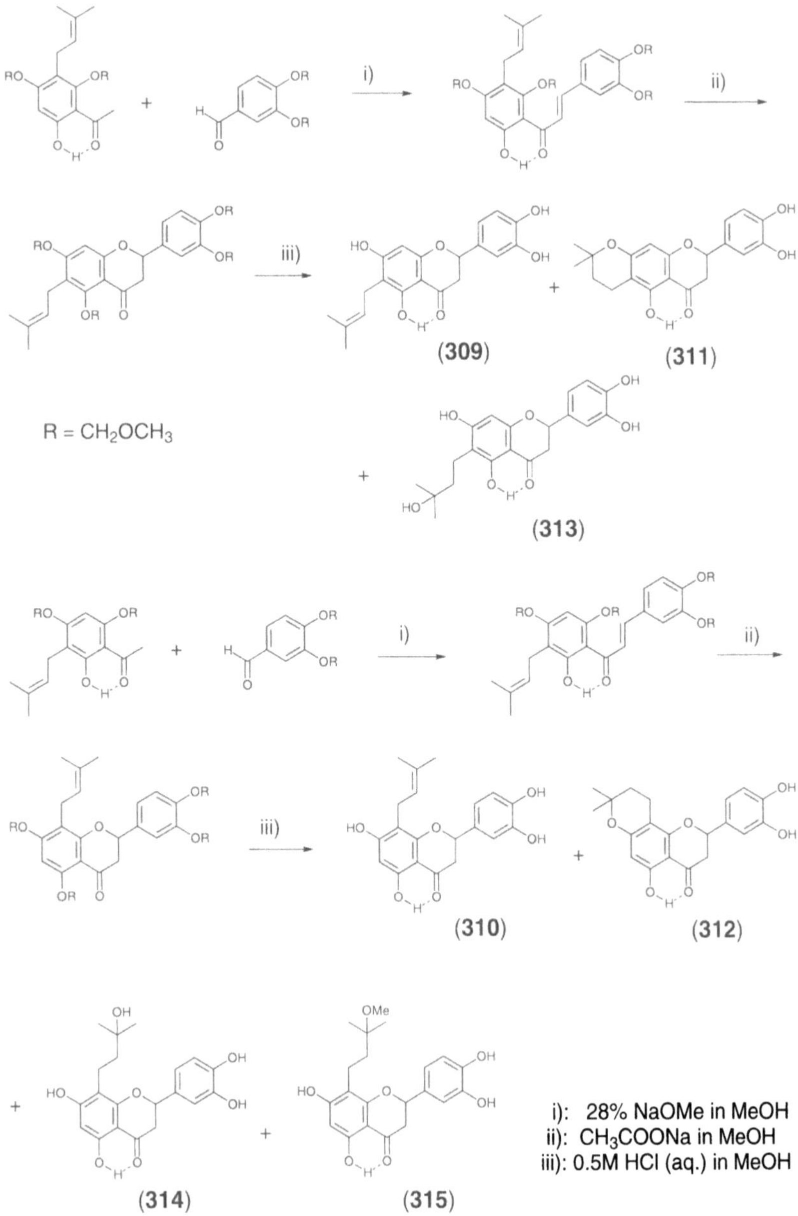

i)
ii)
iii)
(309)
(311)
R = CH2OCH3
(313)
(310)
(312)
(314)
(315)
i): 28% NaOMe in MeOH
ii): CH3COONa in MeOH
iii): 0.5M HCl (aq.) in MeOH

Scheme 5

result that their chemical shifts now obeyed the rule (*410–412*). The remaining two non-conforming monoprenylated flavanones were the 6- and 8-prenylated eriodictyols **309** and **310** isolated from *Wyethia helenioides* (*413*) whose structures were assigned by the cyclization reaction with *p*-toluenesulfonic acid in benzene shown Scheme 4. The chemical shifts of the 5-OH signals of **309** and **310** were the reverse of these predicted by the generalization. Synthesis of 6- and 8-prenylated eriodictyol (**309** and **310**) by the unambiguous method shown in Scheme 5 (*409*) showed that formulas **309** and **310** proposed previously for the natural flavanones had to be reversed.

We also re-investigated the cyclization reactions of **309** and **310**. Both prenylflavanones gave dihydropyranoflavanones **311** and **312** but no **312′** (Fig. 25). The substituent effect on the 5-OH signal of synthetic flavanones (**309**, **310**, **313**, **314** and **315**) agrees with the rule for the other prenylated flavanones as shown in Fig. 25. The difference in the chemical shift of such isomers, *i.e.*, flavanones with a prenyl group at the 6- or 8-position, is considerable (0.32–0.39 ppm). Thus, the chemical

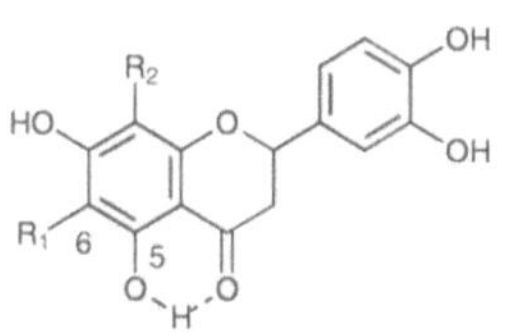

(**309**): R_1 = prenyl, R_2 = H
(**310**): R_1 = H, R_2 = prenyl
eriodictyol (**316**): R_1 = R_2 = H

Chemical Shifts of 5-OH in Acetone-d_6

	5-OH	Δ(**316**) – (**309** - **315**)
(**309**)	12.47	– 0.29
(**313**)	12.48	– 0.30
(**316**)	12.18	-----
(**310**)	12.14	+ 0.04
(**314**)	12.11	+ 0.07
(**315**)	12.12	+ 0.06
(**311**)	12.55	– 0.37
(**312**)	11.89	+ 0.29

309

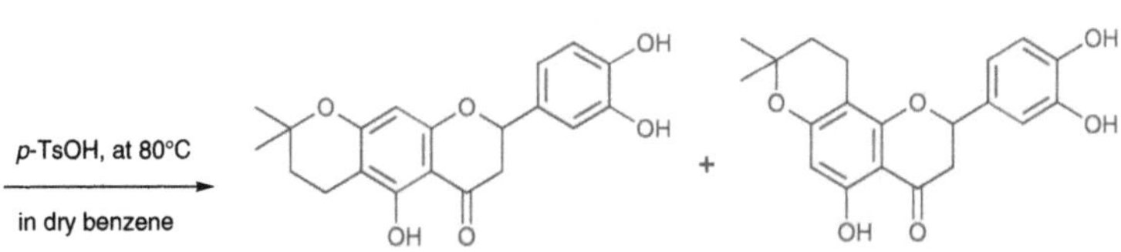

(**311**) (53%) (**312**) (~ 3%)

310 → (*p*-TsOH, at 80°C, in dry benzene) **311** + **312**
(12%) (72%)

(*p*-TsOH: *para*-toluenesulfonic acid)

Fig. 25

shift is a useful tool for differentiating between 6- or 8-prenylated 5,7-dihydroxyflavanones.

The prenylation effect of a 7-*O*-methyl or 7-*O*-prenyl ether of a flavanone is different from that of 7-hydroxyflavanones (*410*). The 5-OH of a 7-*O*-methyl or prenyl ether of a 6-prenylflavanone appears further downfield than that of a 6-unsubstituted flavanone having the same B ring (0.13–0.14 ppm in acetone-d_6, 0.03–0.05 ppm in $CDCl_3$). The -OH signal of an 8-prenylated 7-*O*-methyl or 7-*O*-prenyl ether of a flavanone is also downfield compared with that of an 8-unsubstituted flavanone having the same B ring (0.10–0.11 ppm in acetone-d_6 or $CDCl_3$).

It should be noted that the 5-OH chemical shift depends to some degree on the frequency of the spectrometer used for the measurement (*285*). When the spectrum is measured on a 100 MHz spectrometer, the signal appears further upfield (about 0.1 ppm) than when a 400 MHz instrument is used. The signals of the 5-OH of 6- or 8-prenylated 7-*O*-methylflavanones appeared in a relatively narrow range (δ 12.20–12.24 in acetone-d_6, 12.04–12.12 in $CDCl_3$). Thus, comparison of the 5-OH signal of a 7-*O*-methyl or prenyl ether needs measurement under same conditions, frequency of instrument and used of $CDCl_3$. The chemical shift of the signal also shows a slight but negligible upfield shift as temperature is increased (0.002 ppm per degree) (*285*). The shift with change in concentration is also negligible and does not changed in a concentration range of 0.03–2%.

Table 5. *Substituent Parameters (ppm) on the 5-Hydroxyl Group of Flavanones (in Acetone-d_6)*

		Substituent parameter
(i)	Prenylation at C-6	+0.28 ~ +0.30
	(prenylations at both C-6 and C-8)	
(ii)	Prenylation at C-8	−0.03 ~ −0.06
(iii)	Methylation of 7-OH[a]	[−0.04 (−0.02)[b]][c]
	(a) (6-Prenylated flavanone)[a]	−0.19 (−0.34)[b]
	(b) (8-Prenylated flavanone)[a]	+0.10 (+0.13)[b]
(iv)	Hydroxylation at C-2′	
	(a) (3′-Unsubstituted flavanone)	+0.01 ~ +0.05
	(b) (3′-Alkenylated flavanone)	+0.02
(v)	Hydroxylation at C-3	−0.46 ~ −0.47
(vi)	Hydroxylation at C-3′	+0.01
(vii)	Hydroxylation at C-4′	+0.02 ~ +/0.03

[a] Only one example; [b] Measured in $CDCl_3$; [c] 6,8-Unsubstituted flavanone

Table 6. *Substituent Parameters (ppm) on the 5-Hydroxyl Group of Flavones (in Acetone-d_6)*

	Substituent parameter
(i) Prenylation at C-6	$+0.25 \sim +0.27$
(ii) Prenylation at C-8	$-0.05 \sim -0.07$
(iii) Prenylation at C-3	$\pm 0 \sim +0.03$
(iv) Methylation of 7-OH	
(6-prenylflavone)	$+0.12 \sim +0.18$
(v) Prenylation at C-3′	
(a) (2′-Unsubstituted flavone)	$+0.02 \sim +0.03$
(b) (2′-Hydroxyflavone)	-0.04
(vi) Hydroxylation at C-2′	
(a) (3′-Unsubstituted flavone)	$+0.11 \sim +0.13$
(b) (3′-Alkenylated flavone)	$+0.07$
(vii) Hydroxylation at C-3′	-0.01
(Hydroxylation at C-3: flavonol)	$(-0.82 \sim -0.84)$

Other substituent effects are summarized in Table 5. The shift depending on prenylation at C-6 or C-8 of a flavanone is useful for structure determination of prenylated flavanones (parameters i and ii).*

4.2.2. *Flavones*

The 5-OH signal of flavones appears at δ 12.3–12.6 in acetone-d_6 (*414*). The effects of substitution on the hydroxyl signal are shown in Table 6. Parameter (i) of flavones (prenylation at C-6) is smaller than that in flavanones. The effect of hydroxylation at C-2′ of a flavone (parameter vi) is of considerable value in structure determination as it results in a downfield shift of 0.07–0.13 ppm.

Two geranyl substituted flavones, albanins D and E, were isolated from *Morus alba* and had been assigned as 8-geranylflavones by Takasugi *et al.* using UV spectrometry (*383*, *415*) since Sherif *et al.* had reported earlier that 6-alkyl(alkenyl)flavonoids do not exhibit the aluminum-induced shift due to chelation as a result of steric hindrance (*399*). However, the chemical shifts of the 5-OH of the *Morus* flavones indicated that the compounds were actually 6-geranylflavones. Synthesis

* The chemical shifts of the 5-OH signals reported in this section were observed at 400 MHz in acetone-d_6, unless otherwise stated.

of the 6- and 8-geranylflavones showed that albanins D and E were indeed to 6-geranyl-4′,5,7-trihydroxyflavone and 6-geranyl-2′,4′,5,7-tetrahydroxyflavone, respectively (*414*). Authentic 8-geranyl-2′,4′,5,7-tetrahydroxyflavone isolated from *Brosimopsis oblongifolia* (Moraceae) and thought to be identical with albanin E because of the earlier work (*415*) was actually identical with synthetic 8-geranyl-2′,4′,5,7-tetrahydroxyflavone by direct comparison and has been renamed brosimone L (*416*).

The UV spectra of the synthetic 6-geranylflavones showed the usual bathochromatic shift when aluminum chloride was added as well as did those of the 8-geranylflavones. The other 6-prenylflavones (for example, 5,7-dihydroxy-6-prenylflavone, 6-prenyl-4′,5,7-trihydroxyflavone, 6-prenyl-3′,4′,5,7-tetrahydroxyflavone, *etc.*) also did not show the resistance for the aluminum-induced shift except 3,6-dialkylated(alkenylated) flavones.*

4.2.3. Isoflavones

The 5-OH signal of isoflavones appears between δ 12.5 and 13.5 in acetone-d_6 (*285, 417*). The changes in the chemical shift of the 5-OH signal upon hydroxylation at C-2′ of isoflavones as well as the effect of prenylation at C-6 or C-8 were reported briefly by us (*347*), and in detail later by TAHARA *et al.* (*417*). Hydroxylation at C-2′ of a flavone or flavonol causes a downfield shift of the 5-OH signal (Tables 6 and 8); however, the 5-OH signal of 2′-hydroxyisoflavones appears farther upfield than that of 2′-unsubstituted isoflavones or 2′-*O*-alkyl(alkenyl)-isoflavones (0.16 – 0.56 ppm, Table 7).

The utility of these substituent parameters was shown in the structure determination of the minor isoflavones, kanzonols K (**183**) and L (**184**), isolated from *Glycyrrhiza uralensis* (Section 3.1.2). The ^{1}H NMR and UV spectra (absence of an aluminum-induced shift) of kanzonol K (**183**) showed that the structure of the compound was **183** or **317** (Fig. 26). Calculated values using the substituent parameters of Table 7 for the 5-OH signal of **183** and **317** are δ 12.68 and 13.14, respectively, as shown in Fig. 26. The chemical shift of the 5-OH of kanzonol K (δ

* 6-Prenylated flavanones failed to give the aluminum-induced shift in ethanol but showed the bathochromic shift in methanol. However, on a warm day the shift also occurred in 6-prenylated flavanones on addition of aluminum chloride in ethanol.

Table 7. *Substituent Parameters (ppm) on the 5-Hydroxyl Group of Isoflavones (in Acetone-d_6)*

		Substituent parameter
(i)	Prenylation at C-6 (prenylations at both C-6 and C-8)	+0.28
(ii)	Prenylation at C-8	−0.06
(iii)	Methylation of 7-OH	
	(a) (6-Prenylated isoflavone)	−0.12
	(b) (8-Prenylated isoflavone)	+0.09
(iv)	Hydroxylation at C-2′	
	(a) (3′-Unsubstituted isoflavone)	−0.16
	(b) (3′-Alkenylated isoflavone)	−0.52
(v)	Hydroxylation at C-3′	+0.05
(vi)	Methylation of 3′-OH or 4′-OH	−0.02
(vii)	Cyclization between 2′-OH and 3′-prenyl group	−0.56

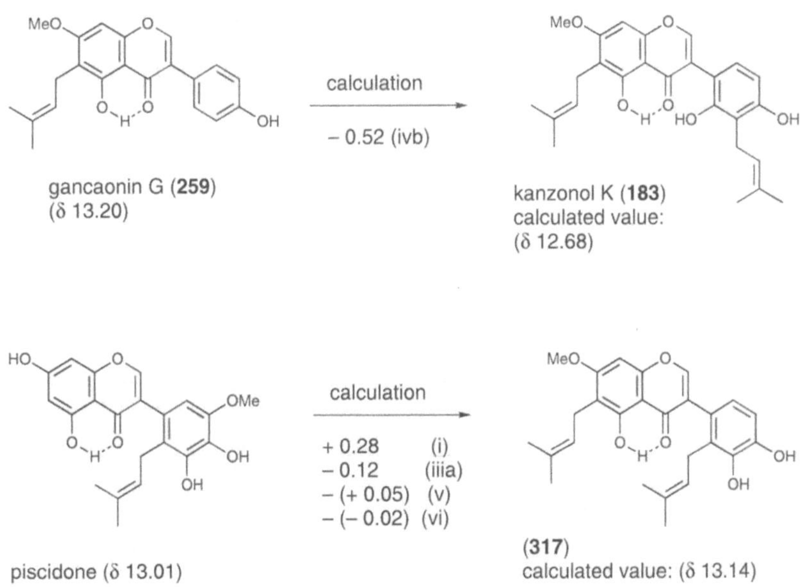

Fig. 26

12.68) agrees with the value calculated for **183**, but not for **317**. Thus, the structure of kanzonol K was **183** (*285*). The structure of kanzonol L (**184**) was also determined in the same manner.

4.2.4. Flavonols

The 5-OH signal of prenylated flavonols (3-hydroxyflavones) is observed between δ 11.9 and 12.6 in acetone-d_6 when the 3-hydroxyl group is free (*276*). The substituent parameters of flavonols are listed in Table 8. It should be noted that the 5-OH signal of flavonols is sometimes shifted due to chelation with metals. When a flavonol fluoresces under irradiation with UV light, the sample needs further purification. The shift upon prenylation at C-6 or C-8 of flavonols (parameters i and ii) is almost identical with that of the other flavonoids. Methylation of 3-OH causes a downfield shift of 0.64–0.67 ppm. However, the 5-OH signal of 3-*O*-methylflavonols appears still further upfield (about 0.2 ppm) than that of flavones having the same A and B rings.

Glepidotin A isolated from the whole plant of *G. lepidota* was reported as an 8-prenylated flavonol (**300**) (Section 3.3). However, the chemical shift of the 5-OH (δ 12.30 in acetone-d_6) indicated that the structure was 5,7-dihydroxy-6-prenylflavonol (**301**); the calculated value for a 6-prenylflavonol (**301**) is δ 12.34–12.35, and the calculated value for an 8-prenylflavonol (**300**) is δ 12.02–12.03. Upon synthesis of 6-prenylflavonol (5-OH, δ 12.35) and 8-prenylflavonol (δ 12.03), glepidotin A was identified as 6-prenylflavonol (**301**) by direct comparison (*276*).

Table 8. *Substituent Parameters (ppm) on the 5-Hydroxyl Group of Flavonols (in Acetone-d_6)*

	Substituent parameter
(i) Prenylation at C-6	+0.25 ~ +0.26
(ii) Prenylation at C-8	−0.06 ~ −0.07
(iii) Methylation of 3-OH	+0.64 ~ +0.67
(iv) Hydroxylation at C-2′	+0.10
(v) Hydroxylation at C-3′	±0 ~ +0.01
(vi) Hydroxylation at C-4′	+0.08 ~ +0.09

4.2.5 Isoflavanones

The 5-OH signal of prenylated isoflavanones appears at δ 12.0–12.7 in acetone-d_6 (*262*). The substituent parameters of isoflavanones are listed in Table 9, but the parameters are incomplete due to the small number of examples in the literature.

Table 9. *Substituent Parameters (ppm) on the 5-Hydroxyl Group of Isoflavanones (in Acetone-d_6)*

		Substituent parameter
(i)	Prenylation at C-6	+0.28 ~ +0.30
(ii)	Prenylation at C-8	−0.04 ~ −0.08
(iii)	Methylation of 7-OH (6-Prenylated isoflavanone)	−0.14 ~ −0.16
(iv)	Hydroxylation at C-2′ (3′-Unsubstituted isoflavanone)	+0.12
(v)	Prenylation at C-3′ (2′,4′-Dihydroxyisoflavanone)	−0.31
(vi)	Methylation of 4′-OH	−0.03 ~ −0.05
(vii)	Cyclization between 2′-OH and 3′-prenyl group	+0.03

Recently, similar substituent effects for hydrogen-bonded hydroxylic protons in 9,10-anthraquinones have been reported by Schripsema and Dagnino (*418*).

4.3. Variations in the Chemical Shift of the 5-Hydroxyl Proton of Pyranoflavonoids

The 5-OH signal of a linear type pyranoflavone such as cudraflavone B (**318**) appears further downfield (0.35 ppm) than that of the comparable flavone with no substituent at the C-6 and C-8 positions (albanin A, **319**) in acetone-d_6 (see Fig. 27) (*414*). In the angular type pyranoflavone morusin (**320**) the -OH signal is also downfield from that of albanin A (0.09 ppm). The notable difference between the chemical shifts of linear and angular type flavones (0.26–0.32 ppm) is a useful tool for structure determination. A similar shift difference is also observed in pyranoisoflavones and pyranoflavanones as shown in Fig. 27 (*417*, *419*). The difference between the chemical shifts of the 5-OH signals in the linear type dihydropyranoflavanone (**311**) and its angular type isomer (**312**) is large (0.66 ppm, Fig. 25). It is also useful for structure determination of pyranoflavanones.

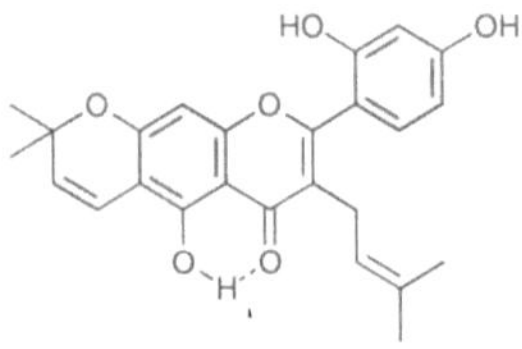

cudraflavone B (318)
(δ 13.51 in acetone-d_6)

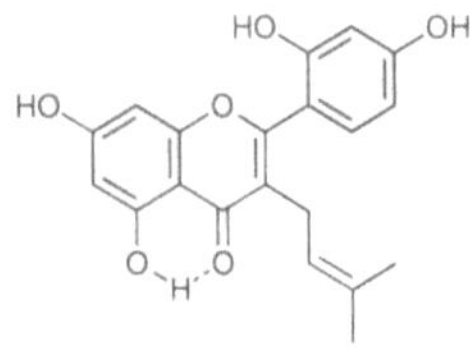

albanin A (319)
(δ 13.16 in acetone-d_6)

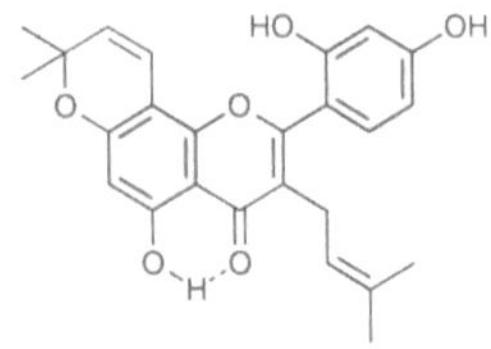

morusin (320)
(δ 13.25 in acetone-d_6)

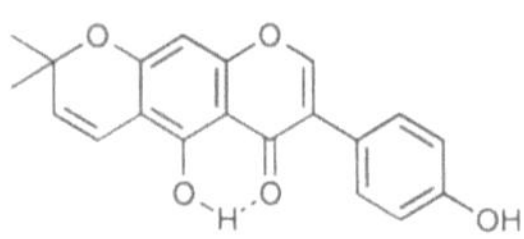

alpinumisoflavone (321)
(δ 13.44 in acetone-d_6)

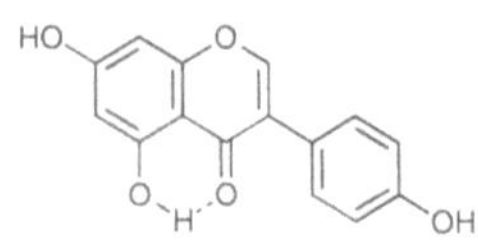

genistein (322)
(δ 13.03 in acetone-d_6)

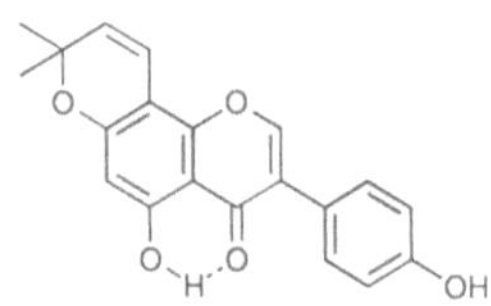

derrone (323)
(δ 13.14 in acetone-d_6)

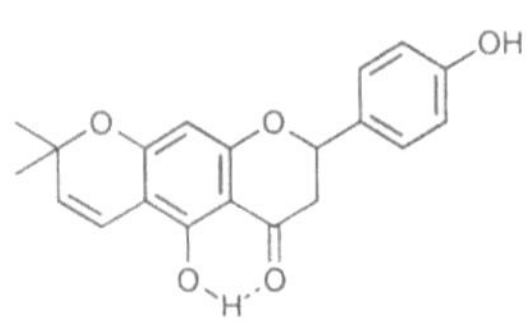

(324)
(δ 12.28 in $CDCl_3$, 270 MHz)

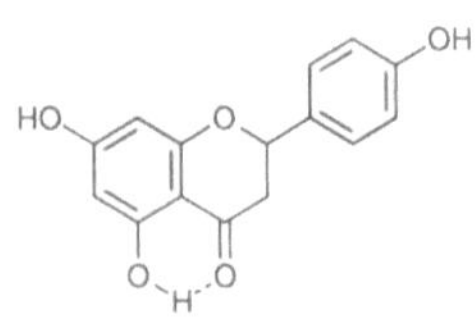

naringenin (325)
(δ 12.03 in $CDCl_3$)

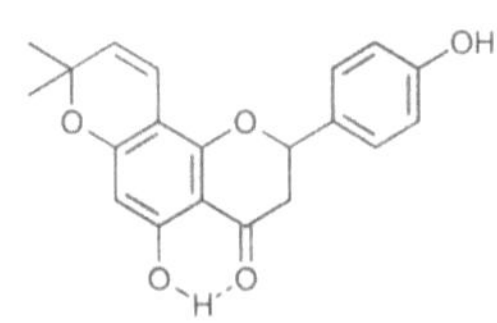

citflavanone (326)
(δ 12.09 in $CDCl_3$, 270 MHz)

Fig. 27

A hypothesis concerning the substituent effects on the 5-OH signal has been presented (*420*). It is reasonable to assume that the hydroxyl group at the C-3 position of a 3-hydroxyflavanone withdraws an electron from the carbonyl oxygen. Thus, the hydrogen bond between the 5-OH and the carbonyl oxygen of 3-hydroxyflavanones is presumably weaker than that of flavanones. Actually the 5-OH signal of 3-hydroxyflavanones appears further upfield than that of flavanones (0.46–0.47 ppm, see

Table 5). This hypothesis was studied using the molecular models calculated by Mopac-6 using PM3 Hamiltonian (Fig. 28). In the model, the electron density at the carbonyl oxygen of 3-hydroxyflavanone (**327**) was smaller than that of flavanone (**328**). Thus the hydrogen bond of 3-hydroxyflavanones is weaker than that of flavanones. On the other hand, no significant difference in electron density (at the 4-carbon and 4-oxygen, 5-carbon, 5-oxygen and 5-hydrogen) was found among the models of 6- or 8-prenylated and non-prenylated flavanone **329**, **267** and **328** (Fig. 29). In the models the distances between the 5-OH hydrogen and the oxygen of the carbonyl group of compounds **329**, **267** and **328** are almost same (Fig. 29).

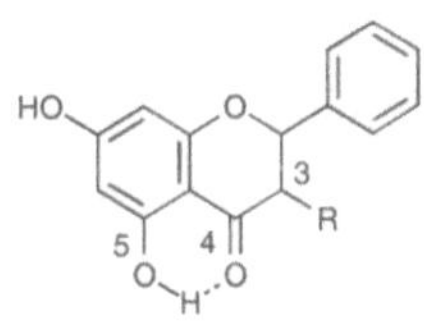

(**327**): R = OH
(**328**): R = H

	Electron density		
	C=O	5-O	5-OH
Flavanonol (**327**)	6.325	6.248	0.750
Flavanone (**328**)	6.361	6.248	0.751

Fig. 28

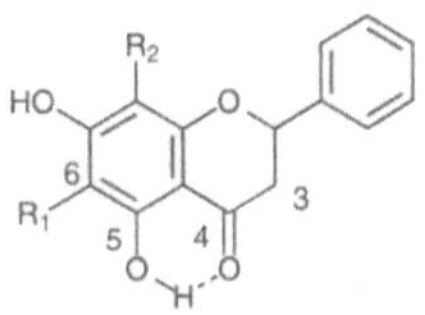

(**328**): $R_1 = R_2 = H$
(**329**): R_1 = prenyl, R_2 = H
(**267**): R_1 = H, R_2 = prenyl

	Distance between 5-OH and 4-CO (Å)
(**328**)	1.83
(**329**)	1.81
(**267**)	1.81

Fig. 29

The signals of the proton of a phenolic hydroxyl and water approach each other as chemical exchange between them increases as, for example, by an increase in the temperature. The chemical exchange is detectable by the saturation transfer method (*421*). Data of saturation transfer between water and the 5-OH of three flavanones are shown in Table 10. The amount of the transfer from the saturated proton of water to the 5-hydroxy proton of 6-prenylated flavanone **329** (0.136H, at 27 °C) was smaller than that in non- and 8-prenylated flavanones **328** (0.398H) and **267** (0.249H)* and is probably due to steric hindrance by the 6-prenyl group. If the downfield shift of the 5-OH of 6-prenylated flavanones is caused by resistance to chemical exchange between the 5-OH proton and water or the other hydroxylic protons of the flavanone, the degree of chemical exchange (saturation transfer, temperature) may affect the chemical shift. However, the change in chemical shift of **329** is small (Table 10, δ 12.429 at 27 °C, δ 12.387 at 55 °C). The amount of the transfer, from saturated proton of water to 5-hydroxyl proton, in **328** (at 27 °C, 39.8%) might be comparable to that in **329** (at 55 °C, 45.7%). But, the chemical shift of the 5-OH of **328** (δ 12.142 at 27 °C) differs from that of the 5-OH of **329** (δ 12.387 at 55 °C). The above data indicate that the contribution of the chemical exchange to the shift of the 5-OH signal accompanying prenylation is small.

Table 10. *Saturation Transfer from H_2O to 5-Hydroxyl Proton of* **267**, **328** *and* **330**

	Earia of 5-OH proton when irradiated at H_2O signal (amount of transfered proton, %)		Chemical shift of 5-OH	
	27 °C	55 °C	27 °C	55 °C
(**328**)	0.602H (39.8)	0.209H (79.1)	12.142	12.090
(**329**)	0.864H (13.6)	0.543H (45.7)	12.429	12.387
(**267**)	0.751H (24.9)	0.340H (66.0)	12.104	12.050

Since the electron densities near the hydrogen bond of **267**, **328** and **329** are about the same and since the contribution of chemical exchange (saturation transfer) is small, the following is the most reasonable cause for the downfield shift of the 5-OH signal on prenylation at the 6-position: The 6-prenyl group interferes with rotation of the bond between the C-5 carbon and the C-5 hydroxyl group and/or distorts the

* Generally, the signal is weaker than the other proton signals when the spectrum is measured using a Fourier transform NMR spectrometer. The relaxation time (T_1) of the 5-OH is between 6–9 seconds at 28 °C in acetone-d_6, 4–6 seconds at 55 °C (unpublished data). When a quantitative spectrum is needed, the pulse delay time must be longer than 10 seconds.

References, pp. 110–140

molecular orbital at the 5-OH hydrogen. Thus the 5-OH signal of 6-prenylated flavonoids may appear further downfield than that of flavonoids having no substituent at C-6 (*422*). The upfield shift of the 5-OH signal, the effect of prenylation at C-8 position, may be caused by distortion of the orbital at the 5-OH proton.

4.4. CD Spectra of Licorice Isoflavans

The CD spectra of polyoxygenated isoflavans exhibit simple Cotton effects at the ^1L_a-band (near 235 nm) and the ^1L_b-band (near 290 nm) while *R*-isoflavans show a negative Cotton effect at the ^1L_a-band and a positive Cotton effect(s) at the ^1L_b-band, *S*-isoflavans show a positive Cotton effect at the ^1L_a-band and negative Cotton effect(s) at the ^1L_b-band (*423*). In the CD spectra of isoprenoid-substituted isoflavans obtained from licorice (Table 11), eighteen compounds (**131**, **147**, **155**, **164**, **165**, **180–182**, **185–187**, **195**, **196**, **224–226**, **265**, **266**) showed a negative Cotton effect at the ^1L_a-band, two compounds (**199** and **264**) showed a positive Cotton effect and three compounds (**126**, **200** and **201**) showed split Cotton effects in the region. Some (14 compounds, **126**, **131**, **147**, **155**, **181**, **182**, **185**, **186**, **199**, **200**, **224**, **225**, **264** and **266**) showed a positive Cotton effect(s) at the ^1L_b-band, and **226** showed two negative Cotton effects in this region. The remaining eight isoflavans (**164**, **165**, **180**, **187**, **195**, **196**, **201** and **265**) showed split Cotton effects at the ^1L_b-band.* No correlation between the CD curves and the substitution patterns of the licorice isoflavans and some of their derivatives exists (Table 11). Probably the changes in the Cotton effect at the ^1L_b-band are due to adjacent overlapping optically active absorption bands as reported in the case of flavanones (*233*). Some isoflavans were correlated chemically, *e.g.*, **164** and **330**, and the CD curves of a few isoflavans changed on presumably unimportant structural alterations, *e.g.*, **126** to **131**, **165** to **181** and **331** (Table 11).

GAFFIELD has reported that the sign of the Cotton effect near 230 nm is always positive for *trans*-2*R*,3*R*-3-hydroxyflavanones and usually negative for 2*S*-flavanones (*233*). This is probably applicable to isoflavan

---

* Previously, we ascribed the split Cotton effect at the ^1L_b-band to chiral exciton coupling between the A and B ring chromophores (*261, 340*). However, the probability of such coupling is very small because the molecular ellipticity value of the ^1L_b-band (for example, in licoricidin (**164**) $[\theta]_{272}$ − 1500, $[\theta]_{288}$ + 2100 and in glyasperin D (**196**) $[\theta]_{274}$ − 1900, $[\theta]_{288}$ + 1100) is not larger than that in simple 3*R*-isoflavans (for example, in 3*R*-vesitol (2′,7-dihydroxy-4′-methoxyisoflavan) $[\theta]_{290}$ + 2220) (*426*).

Table 11. *Cotton Effects of Isoflavan Derivatives in Methanol*

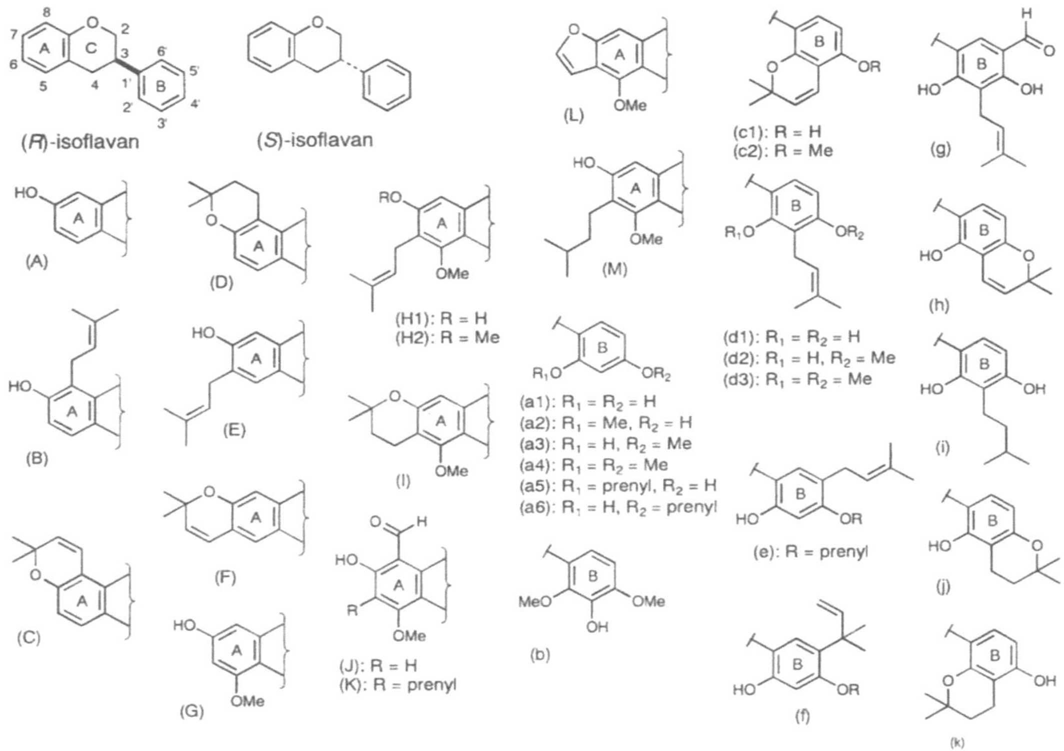

Trivial name	A ring	B ring	Sign of Cotton effect		Reference
			^1L_a band (~ 235 nm)	^1L_b band (270 – 290 nm)	
Glabridin (**126**)	(C)	(a1)	split [+ (224), – (238)]	+ (292)	(*252*)
Hispaglabridin A (**131**)	(C)	(d1)	– (234)	+ (292)	(*a*)
Kanzonol X (**147**)	(B)	(d1)	– (233)	+ (286)	(*252*)
Kanzonol R (**155**)	(G)	(d2)	– (235)	+ (273), + (278)	(*261*)
Licoricidin (**164**)	(H1)	(d1)	– (238)	split [– (272), + (288)]	(*284*)
Licorisoflavan A (**165**)	(H2)	(d1)	– (238)	split [– (275), + (286)]	(*284*)
Kanzonol H (**180**)	(I)	(d1)	– (238)	split [– (273), + (287)]	(*284*)
Kanzonol I (**181**)	(H2)	(c1)	– (237)	+ (287)	(*284*)
Kanzonol J (**182**)	(I)	(c1)	– (237)	+ (277), + (289)	(*284*)
Kanzonol M (**185**)	(J)	(d2)	– (234 sh)	+ (289)	(*261*)
Kanzonol N (**186**)	(J)	(d1)	– (235)	+ (288)	(*261*)
Kanzonol O (**187**)	(J)	(c1)	– (234)	split [+ (277), – (289)]	(*261*)
Glyasperin C (**195**)	(H1)	(a1)	– (239)	split [- (275), + (289)]	(*288*)
Glyasperin D (**196**)	(H2)	(a1)	– (239)	split [– (274), + (288)]	(*288*)
Glyasperin G (**199**)*	(L)	(d1)	+ (233)	+ (276), + (290)	(*290*)
Glyasperin H (**200**)*	(C)	(b)	split [– (226), + (239)]	+ (277)	(*290*)
Glyasperin I (**201**)*	(H1)	(a2)	split [+ (233), – (242)]	split [– (276), + (288)]	(*290*)
Glyinflanin I (**224**)	(E)	(c1)	– (236)	+ (284 sh), + (293)	(*299*)
Glyinflanin J (**225**)	(F)	(c1)	– (236)	+ (278)	(*299*)
Glyinflanin K (**226**)	(C)	(c1)	– (233)	– (284), – (293)	(*299*)
Gancaonin X (**264**)*	(F)	(a3)	+ (228), + (244 infl.)	+ (274)	(*340*)
Gancaonin Y (**265**)	(A)	(c2)	– (232)	split [– (272), + (291)]	(*340*)
Gancaonin Z (**266**)	(A)	(d2)	– (231)	+ (285), + (289 sh)	(*340*)
Dihydroglabridin (**332**)	(D)	(a1)	– (235)	+ (285 sh), + (289)	(ud)
No name	(C)	(a5)	split [+ (227), – (239)]	+ (292)	(*a*)

Table 11 (*continued*)

Trivial name	A ring	B ring	Sign of Cotton effect		Reference
			^1L_a band (~ 235 nm)	^1L_b band (270 – 290 nm)	
No name	(C)	(a6)	split [+ (227), – (241)]	+ (291)	(*a*)
No name	(C)	(e)	split [+ (228), – (242)]	+ (294)	(*a*)
No name	(C)	(f)	split [+ (228), – (242)]	+ (293)	(*a*)
Tetrahydrolicoricidin (**330**)	(M)	(i)	– (235)	split [– (272), + (286)]	(ud)
Licoricidin trimethyl ether	(H2)	(d3)	– (236)	split [– (271), + (289)]	(ud)
Isolicoricidin A (**333**)	(I)	(j)	split [+ (234), – (246)]	split [– (273), + (288)]	(*b*)
Isolicoricidin B (**334**)	(I)	(k)	split [+ (233), – (245)]	split [– (272), + (288)]	(*b*)
No name	(H1)	(g)	– (237 sh)	– (282), – (310)	(*c*)
No name	(K)	(a1)	– (236)	– (280), + (300 sh)	(*c*)
No name (**331**)	(H2)	(h)	– (239 sh)	– (280)	(*284*)

* The absolute configuration of the compounds could not be determined by CD spectra. The other compounds were assigned as *R*-isoflavan by their CD spectrum or chemical correlation; (*a*): derived from glabridin (**126**) with the prenylation reaction; (*b*): derived from licoricidin (**165**) according to the previous literature (*267*); (*c*): derived from licoricidin (**164**) with Friedel-Crafts reaction (unpublished data)

derivatives as well. On this basis the eighteen isoflavans, which showed a negative Cotton effect at the ^1L_a-band, were assigned the *R*-isoflavan configuration (*424*). The stereochemistry of the remainder (**126**, **199**, **200**, **201**, **264**) could not be assigned on the basis of their CD spectra.

Glabridin (**126**) has been assigned the *R*-isoflavan configuration because of its CD spectrum which exhibits a positive Cotton effect at 300 nm (*250*). It showed split Cotton effects at the ^1L_a-band, but a prenylated compound derived from **126**, hispaglabridin A (**131**), showed a negative Cotton effect at the ^1L_a-band (Table 11). Dihydroglabridin (**332**), derived from **126**, also showed a negative Cotton effect at the ^1L_a-band. On the other hand, licoricidin (**164**) exhibited a negative Cotton effect at the ^1L_a-band, but isolicoricidins A and B (**333** and **334**) derived from **164** showed split Cotton effects at the ^1L_a-band. The signs of the split Cotton effects of **333** and **334** are the same as those of **126**. Thus, the earlier assignment of **126** as an *R*-isoflavan may be correct.

The Cotton effect at the ^1L_a-band of glyasperin G (**199**) and gancaonin X (**264**) may overlap with adjacent strong Cotton effects; for **199** $[\theta]_{228} + 11200$, $[\theta]_{233} + 9200$ (sh), $[\theta]_{240} + 3400$ (valley), $[\theta]_{254} + 6200$, and for **264** $[\theta]_{228} + 12500$, $[\theta]_{244} + 3400$ (sh). Elucidation of the stereochemistry of some isoflavans obtained from licorice such as glyasperins G, H and I (**199**, **200** and **201**) and gancaonin X (**264**) needs further work.

Kurosawa *et al.* have reported that the ORD curves and NMR spectra of isoflavans are markedly influenced by which conformations are preferred in solution (*425*). From ^1H NMR measurements (in CD_3OD), all isoflavans obtained from *Glycyrrhiza* species have the conformation shown in Fig. 30, namely the A-form. A conformational isomer (**335**) of glabridin (**126**) was obtained by isomerization of **126** in basic solution (Fig. 30). Isomer **335** has the B conformation; the CD spectrum was quite different from that of **126** (Fig. 30), and the sign of the Cotton effect at the ^1L_a-band was negative (unpublished data).

5. Biological Activities of Phenolic Constituents of *Glycyrrhiza* Species

Okuda *et al.* have reported that some licorice flavonoids, isolicoflavonol (**171**), glycycoumarin (**172**), licochalcones A and B (**213** and **214**), glycyrrhisoflavone (**250**) and licopyranocoumarin (**256**), inhibit the cytopathic activity of HIV (*427*). Shibata *et al.* have reported that licochalcone A (**213**) shows anti-tumor promoting activity in a two-stage mouse skin carcinogenesis experiment (*428*). Inhibitory activity of

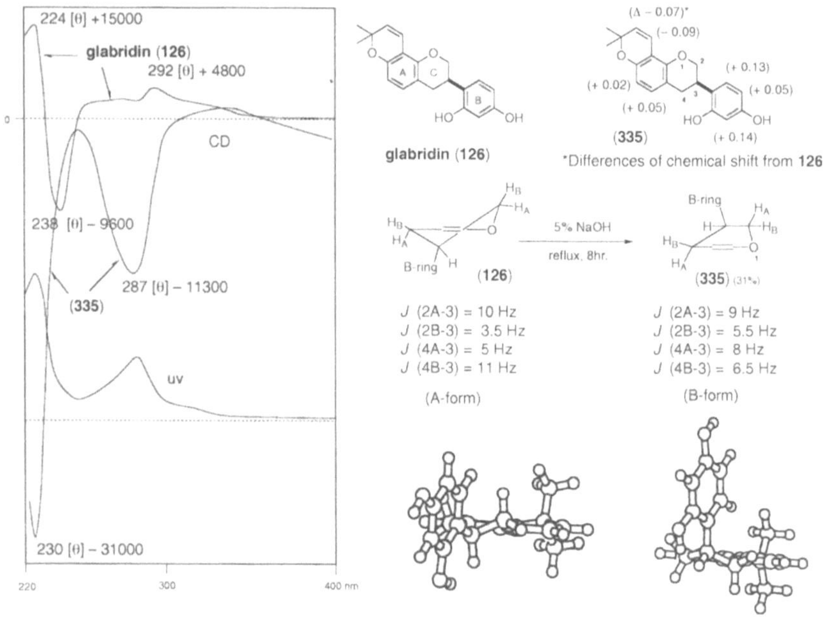

Fig. 30

compound (**213**) against growth of the human pathogenic protozoans, *Leishmania major* and *L. donovani*, has been reported by KHARAZMI *et al.* (*429*). They have also reported that pyranochalcone (**231**) and kanzonol C (**238**) inhibit the growth of *L. donovani* (*305*). Kanzonol C (**238**) has previously been synthesized in an attempt to develop anti-ulcer drugs and exhibits an anti-ulcer effect on both Shay's pylorus-ligated rats and water-immersed and restraint stress rats (*430*).

KAN *et al.* have reported the effect of a pterocarpan, homopterocarpin (Table 13), isolated from *G. pallidiflora*, on human throat cancer cells, HEp-2 (*431*). The compound inhibits multiplication (at 1 μmol / l) or kills the cells at higher concentrations. MITSCHER *et al.* described the antimutagenic activity of flavonoids obtained from *G. glabra*, in which glabrene (**130**) shows high activity (*432*). The radical scavenging effect of licochalcone B (**214**) has been reported by OKUDA *et al.* (*335*). KIMURA *et al.* have reported the effect of licochalcones A and B (**213** and **214**) on leukotriene formation in human polymorphonuclear neutrophils (*433*).

MITSCHER *et al.* have reported antimicrobial activities of flavonoids obtained from *G. glabra*, glabridin (**126**), glabrol (**127**), hispaglabridins A and B (**131** and **132**), 4′-*O*-methylglabridin (**133**) and 3-hydroxyglabrol (**135**), against *Staphylococcus aureus* or *Mycobacterium smegmatis* (*254*). They have also reported the antimicrobial activity of flavonoids obtained from *G. lepidota*, but the potency is not particularly high (*360, 361*). NAMBA *et al.* reported that isoglycyrol (**168**), glycyrol (**169**), glycycoumarin (**172**) and glycyrin (**248**) isolated from *G. uralensis* have potent antibacterial action against a cariogenic bacterium, *Streptococcus mutans* (*330*). Antimicrobial activity of some licorice flavonoids was also reported by TABATA *et al.* (*341*) and DEMIZU and OKADA *et al.* (*332*, *434*).

Antiinflammatory activity of pinocembrin (**328**) obtained from aerial parts of *G. glabra* has been reported by VAKHABOV *et al.* (*435*). KOBAYASHI *et al.* have reported an inhibitory effect of isoliquiritin (**109**), liquiritigenin (**111**) and isoliquiritigenin (**112**) on tube formation from vascular endothelial cells of rats (*436*). Compound (**112**) also exhibits a preventive effect on ulcer formation induced by severe necrotizing agents in rats (*437*) and anti-platelet action (*438*): Anti-platelet action, inhibition of platelet aggregation, phosphorylation of 40K and 20K dalton proteins, inositol 1,4,5-trisphosphate production, intraplatelet calcium increase and phosphodiesterase activity *in vitro*, of licopyranocoumarin (**256**) have been reported by TAWATA *et al.* (*439*). Licorisoflavan A (**165**) has been isolated as a potent benzodiazepin-binding stimulator by LAM *et al.* (*273*, *440*).

There are some reports which describe screening of biologically active compounds from licorice by use of enzymes, *e.g.*, xanthine oxidase (*228*, *337*), monoamine oxidase (*228*, *338*, *441*), aldose reductase (*442*), phosphodiesterase (*443*), hyaluronidase (*444*), etc. In our screening of biological active compounds by using enzymes, some compounds exhibited inhibitory activity (*445*, *446*): Thus gancaonins E (**278**, IC_{50}:1.4×10^{-8} mol/l), R (**285**, 1.1×10^{-8} mol/l), S (**286**, 7.8×10^{-8} mol/l) and U (**288**, 1.1×10^{-8} mol/l) showed inhibitory activity for 5-lipoxygenase (positive control; cirsiliol, IC_{50}: 1.0×10^{-7} mol/l). Gancaonin S (**286**, IC_{50}, 2.1×10^{-5} mol/l) exhibited inhibited cyclooxygenase (positive control: indomethacin, IC_{50}: 1.5×10^{-6} mol/l). The following flavonoids also exhibited inhibitory activity, but the potency was not particularly high (*447*): licoricidin (**164**), glycyrol (**169**), gancaonins A (**274**), B (**275**), F (**258**) and J (**244**) for 5-lipoxygenase, gancaonins A (**274**), B (**275**), R (**285**), S (**286**) and U (**288**) for cyclooxygenase, licoricidin (**164**), gancaonins J (**244**), R (**285**), S (**286**) and U (**288**) for Na^+, K^+-ATPase, isoglycyrol (**168**), gancaonins E (**278**), R (**285**) and U (**288**) for aldose reductase, gancaonin J (**244**) for testosterone 5α-reductase. Among the above compounds, flavonoids having relatively high activity have been isolated from the aerial parts of licorice or from *Glycyrrhiza* species not used as drugs, but not clinically used licorice (*447*).

Table 12. *Physical Data and Plant Sources of Saponins*

Trivial name	$[\alpha]_D$	mp[a]	Plant source	Ref.
$C_{41}H_{62}O_{14}$				
Apioglycyrrhizin (**72**)	+43° (MeOH)	193–195 °C	*G. inflata*	(*209*, *210*)
			G. uralensis	(*206*)
			G. glabra (Chinese)	(*260*)
Araboglycyrrhizin (**73**)	+31° (MeOH)	225–230 °C	*G. inflata*	(*209*)
	+31° (MeOH)	237–238 °C	*G. inflata*	(*210*)
			G. uralensis	(*206*)
			G. glabra (Chinese)	(*260*)
$C_{42}H_{60}O_{16}$				
Licorice-saponin E2 (**62**)	+68° (MeOH)	216–219 °C	*G. uralensis*	(*203*)
	+68° (MeOH)	218–219 °C	*G. uralensis*	(*205*, *206*)
			G. glabra (Chinese)	(*260*)
Yunganoside E2 (**88**)	−42° (MeOH)	powder	*G. yunnanensis*	(*217*)
$C_{42}H_{62}O_{15}$				
Licorice-saponin C2 (**60**)	−120° (MeOH)	249–251 °C	*G. uralensis*	(*100*, *203*)
			G. glabra (Chinese)	(*260*)
Yunnanglysaponin A (**82**)		178°C (dec.)	*G. yunnanensis*	(*216*)
$C_{42}H_{62}O_{16}$				
Glycyrrhizic acid (**2**)		214 °C	*G. aspera*	(*98*)
		219–220 °C	*G. inflata*	(*448*)
			G. inflata	(*210*, *451*)
			G. uralensis	(*99*, *100*, *206*)
	+55.8° (MeOH)	217 °C	*G. uralensis*	(*127*)
			G. glabra (Chinese)	(*260*)
			G. korshinskyi	(*99*)
			G. glabra (HPLC)	(*36*, *225*, *450*)
			G. inflata (HPLC)	(*36*, *225*)

			G. aspera (HPLC)	(*36*)
			G. eurycarpa (HPLC)	(*36, 225*)
Uralsaponin A (**2**)	+ 42.7° (MeOH)	235 °C	*G. uralensis*	(*127*)
	+ 43.1° (MeOH)	212–215 °C	*G. eurycarpa*	(*128*)
Licorice-saponin H2 (**65**)	+ 31.3° (MeOH)	209–210 °C	*G. uralensis*	(*204, 205, 206*)
			G. inflata	(*209, 210*)
			G. glabra (Chinese)	(*260*)
Licorice-saponin K2 (**67**)	+ 28.0° (MeOH)	207–209 °C	*G. uralensis*	(*204, 205*)
18α-Glycyrrhizin (**69**)			*G. uralensis*	(*206*)
			G. glabra (HPLC)	(*450*)
Uralsaponin B (**70**)	+ 31.0° (MeOH)	244 °C (dec.)	*G. uralensis*	(*127*)
	+ 36.2° (MeOH)	244 °C (dec.)	*G. eurycarpa*	(*208*)
		240 °C (dec.)	*G. inflata*	(*451*)
			G. eurycarpa (HPLC)	(*36, 225*)
			G. uralensis (HPLC)	(*36, 225*)
			G. inflata (HPLC)	(*36, 225*)
			G. glabra (HPLC)	(*36, 225*)
Glyeurysaponin (**71**)	+ 22.5° (MeOH)	288 °C (dec.)	*G. eurycarpa*	(*208*)
Glyyunnanpro-sapogenin D (**81**)		168–170 °C (as hexaacetyl dimethyl ester)	*G. yunnanensis*	(*194*)
Yunnanglysaponin B (**83**)		219 °C (dec.)	*G. yunnanensis*	(*216*)
Yunganoside F2 (**89**)	−30° (MeOH)	powder	*G. yunnanensis*	(*217*)
Yunganoside J2 (**97**)	+170° (MeOH)	powder	*G. yunnanensis*	(*218*)
Yunganoside L2 (**101**)	+154° (MeOH)	powder	*G. yunnanensis*	(*218*)
Macedonic acid diglucronide (**102**)	−33.1°	218–220 °C	*G. echinata*	(*219*)
			G. pallidiflora	(*219*)
$C_{42}H_{62}O_{17}$				
Licorice-saponin G2 (**64**)	+34.0° (MeOH)	229–230 °C	*G. uralensis*	(*204, 205, 206*)
			G. inflata	(*209, 210*)
			G. glabra (Chinese)	(*260*)

Table 12 (*continued*)

Trivial name	$[\alpha]_D$	mp[a]	Plant source	Ref.
Yunganoside K2 (**99**)	−30.4° (MeOH)	powder	*G. yunnanensis*	(*218*)
$\mathbf{C_{42}H_{64}O_{15}}$				
Licorice-saponin B2 (**59**)	+54° (MeOH)	209–210 °C	*G. uralensis*	(*100*, *203*)
Yunganoside H2 (**93**)	+125.7° (MeOH)	powder	*G. yunnanensis*	(*218*)
Yunganoside I2 (**95**)	+206.0° (MeOH)	powder	*G. yunnanensis*	(*218*)
$\mathbf{C_{42}H_{64}O_{16}}$				
Licorice-saponin J2 (**66**)	+21.0° (MeOH)	263–265 °	*G. uralensis*	(*204*, *205*)
$\mathbf{C_{42}H_{64}O_{17}}$				
Yunganoside G2 (**91**)	−27.2° (MeOH)	powder	*G. yunnanensis*	(*218*)
$\mathbf{C_{46}H_{76}O_{21}}$				
Yunganoside C1 (**86**)	−10° (MeOH)	powder	*G. yunnanensis*	(*217*)
$\mathbf{C_{47}H_{72}O_{16}}$				
Soyasaponin II (**4**)			*G. glabra* (calls)	(*138*)
Inflasaponin I (**74**)		256–258 °C	*G. inflata*	(*211*, *213*)
$\mathbf{C_{47}H_{76}O_{17}}$				
Licorice-saponin F3 (**63**)	−20° (MeOH)	215–217 °C	*G. uralensis*	(*204*)
	−20.0° (MeOH)	214–217 °C	*G. uralensis*	(*205*)
$\mathbf{C_{48}H_{72}O_{20}}$				
Yunganoside J1 (**96**)	+143.2° (MeOH)	powder	*G. yunnanensis*	(*218*)
Yunganoside L1 (**100**)	+180° (MeOH)	powder	*G. yunnanensis*	(*218*)
$\mathbf{C_{48}H_{72}O_{21}}$				
Licorice-saponin A3 (**58**)	+69° (MeOH)	198–199 °C	*G. uralensis*	(*100*, *206*)
	+69° (MeOH)	196–199 °C	*G. uralensis*	(*203*)
			G. inflata	(*209*, *210*)
			G. glabra (Chinese)	(*260*)
Yunganoside K1 (**98**)	−10.7° (MeOH)	powder	*G. yunnanensis*	(*218*)
$\mathbf{C_{48}H_{74}O_{16}}$				

Inflasaponin II (**75a**)		178–180 °C	*G. inflata*	(*211, 214*)
Inflasaponin III (**76**)		271–272 °C	*G. inflata*	(*211, 212*)
Inflasaponin VI (**75b**)		197–199 °C	*G. inflata*	(*214*)
$C_{48}H_{74}O_{19}$				
Yunganoside H1 (**92**)	+95.7° (MeOH)	powder	*G. yunnanensis*	(*218*)
Yunganoside I1 (**94**)	+180° (MeOH)	powder	*G. yunnanensis*	(*218*)
$C_{48}H_{74}O_{21}$				
Yunganoside D1 (**87**)	+11° (MeOH)	powder	*G. yunnanensis*	(*217*)
Yunganoside G1 (**90**)	−7.4° (MeOH)	powder	*G. yunnanensis*	(*218*)
$C_{48}H_{76}O_{21}$				
Yunganoside A1 (**84**)	+6° (MeOH)	powder	*G. yunnanensis*	(*217*)
Yunganoside B1 (**85**)	+8° (MeOH)	powder	*G. yunnanensis*	(*217*)
$C_{48}H_{78}O_{18}$				
Soyasaponin I (**3**)			*G. glabra* (callus)	(*138*)
$C_{49}H_{76}O_{20}$				
Licorice-saponin L3 (**68**)	+3.7° (MeOH)	233–234 °C	*G. uralensis*	(*206*)
$C_{50}H_{78}O_{16}$				
Inflasaponin IV (**77**)		234–236 °C	*G. inflata*	(*211, 213*)
Inflasaponin V (**78**)		291–291 °C	*G. inflata*	(*211, 212*)
$C_{50}H_{76}O_{21}$				
Licorice-saponin D3 (**61**)	−5.0° (MeOH)	powder	*G. uralensis*	(*203, 205*)
$C_{54}H_{86}O_{23}$				
Glabranin A (**79**)	+60° (MeOH)	160–163 °C	*G. glabra*	(*215*)
$C_{59}H_{94}O_{27}$				
Glabranin B (**80**)	+72° (MeOH)	165–167 °C	*G. glabra*	(*215*)

[a] (dec.) means decomposition

Table 13. *Structures and Plant Sources of Phenolic Compounds* (the structure shows in parenthesis as follows: isoflavone (3′,5,7–3 × OH, 6-prenyl, Dmp[5′,4′]) means 2″,2″-dimethylpyrano[5″,6″:5′,4′]-6-prenyl-3′,5,7-trihydroxyisoflavone (**260**); HPLC = detection with HPLC analysis; ud = our unpublished data; n.r. = not reported; callus means cell cultured of *Glycyrrhiza* species)

Chalcone

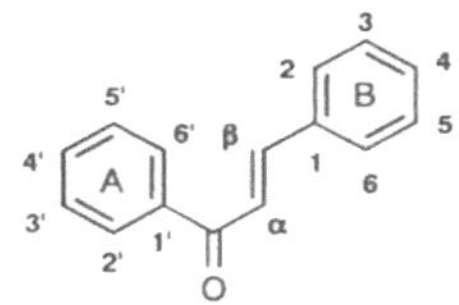

Trivial name	Plant sources	Plant part(s)	References
2′,4′-Dihydroxychalcone	*G. astragalina*	aerial part	(*452*)
4′-Methoxy-4-hydroxychalcone	*G. yunnanensis*	n.r.	(*453*)
Isobavachalcone (2′,4,4′-3×OH, 3′-Prenyl)	*G. uralensis* (callus)		(*454*)
Paratocarpin B (2′,4-2×OH, 3-Prenyl, Dmp[3′,4′])	*G. inflata*	root	(ud)
Isoliquiritin (**109**)	*G. glabra*	root	(*231, 232a, 236, 237, 240*)
	G. glabra (Xinjiang)	root	(*260*)
	G. glabra (Xinjiang, HPLC)	root	(*36, 379, 380*)
	G. glabra (HPLC)	root	(*228, 239*)
	G. uralensis	root	(*206, 232a, 266, 442a*)
	G. uralensis (HPLC)	root	(*36, 228, 326, 379, 380*)
	G. aspera	root	(*98*)
	G. aspera (HPLC)	root	(*36*)
	G. inflata	root	(*433*)
	G. inflata	n.r.	(*451*)
	G. inflata (HPLC)	root	(*36, 228, 379, 380*)

	G. eurycarpa	root	(*308, 309*)
	G. eurycarpa (HPLC)	root	(*36, 380*)
	G. pallidiflora (HPLC)	root	(*36*)
	G. yunnanensis (HPLC)	root	(*36*)
	G. squamulosa (HPLC)	root	(*36*)
	Northwest licorice (HPLC)		(*326, 329*)
Isoliquiritigenin (**112**)	*G. glabra*	root	(*232a, 234, 235, 240*)
	G. glabra	aerial part	(*342*)
	G. glabra (HPLC)	root	(*228, 239*)
	G. glabra (Xinjiang, HPLC)	root	(*36, 379*)
	G. uralensis	root	(*232a, 281, 443*)
	G. uralensis	n.r.	(*241*)
	G. uralensis (HPLC)	root	(*36, 228, 379*)
	G. aspera (HPLC)	root	(*36*)
	G. inflata	root	(*211, 299, 433, 448*)
	G. inflata	n.r.	(*451*)
	G. inflata (HPLC)	root	(*36, 228, 379*)
	G. pallidiflora	root	(*78, 312, 315, 455*)
	G. squamulosa	root	(*456*)
	G. squamulosa (HPLC)	root	(*36*)
	G. eurycarpa (HPLC)	root	(*36*)
	G. yunnanensis (HPLC)	root	(*36*)
	Xinjiang licorice		(*287*)
	Tiexin gancao		(*340*)
	G. uralensis (callus)		(*454*)
Neoisoliquiritin (**114**)	*G. glabra*	root	(*232a, 236*)
	G. glabra (Xinjiang)	root	(*260*)
	G. uralensis	root	(*206, 232a, 266*)
	G. uralensis	n.r.	(*241*)
	Northwest licorice (HPLC)		(*329*)

Table 13 (*continued*)

Trivial name	Plant sources	Plant part(s)	References
Rhamno-isoliquiritin (**118**)	*G. glabra*	root	(*240*)
	G. uralensis	root	(*443*)
Licuraside (**119**)	*G. glabra*	root	(*237*)
	G. glabra (HPLC)	root	(*239*)
	G. uralensis	root	(*442a*)
(Neolicuroside)	*G. glabra*	root	(*243*)
(Isoliquiritin apioside)	*G. glabra* (Xinjiang)	root	(*260*)
	G. uralensis	root	(*100*, *206*)
	G. inflata	root	(*210*)
	G. eurycarpa	root	(*128*, *309*)
Licurazid (**120**)	*G. glabra*	root	(*234a*, *474*)
	G. uralensis	root	(*232a*)
	G. uralensis	n.r.	(*241*)
(Liquiraside, Licuraside)[a]	*G. glabra*	root	(*242*)
(Licuroside)	*G. glabra*	root	(*243*)
(Licraside)	*G. glabra*	root	(*206*)
	G. uralensis	root	(*443*)
Isoliquiritin apioside monoferuloyl ester (**159**)	*G. uralensis*	root	(265)
No name (**190**)	*G. uralensis*	root	(*286*)
Licochalcone A (**213**)	*G. glabra* (Xinjiang)	root	(*260*, *263*)
	G. glabra (Xinjiang, HPLC)	root	(*36*, *379*)
	G. uralensis (Xinjiang, HPLC)	root	(*379*, *380*)
	G. inflata	root	(*295*, *304*, *433*, *434*, *448*, *457*)
	G. inflata (HPLC)	root	(*36*, *228*, *379*, *380*)
	G. eurycarpa	root	(*294*)
	G. eurycarpa (HPLC)	root	(*36*, *380*)
	G. yunnanensis (HPLC)	root	(*36*)
	Xinjiang licorice		(*296*, *335*)
	Tiexin gancao		(*340*)
(Licochalcone)	*G. inflata* (synthesis)	root	(*298*)

Licochalcone B (**214**)	*G. glabra* (Xinjiang)	root	(*260*)
	G. inflata	root	(*304*, *433*, *434*, *448*)
	G. inflata (HPLC)	root	(*228*)
	Xinjiang licorice		(*296*, *335*)
	Tiexin gancao		(*340*)
Glyinflanin G (**222**)	*G. glabra* (Japanese)	root	(*252*)
	G. inflata	root	(*299*)
Licochalcone C (**229**)	*G. inflata*	root	(*304*)
Licochalcone D (**230**)	*G. inflata*	root	(*304*)
No name (**231**)	*G. glabra* (Japanese)	root	(*252*)
	G. glabra	root	(*257*)
	G. inflata	root	(*305*)
Kanzonol B (**237**)	*G. eurycarpa*	root	(*294*)
	Tiexin gancao		(*340*)
Kanzonol C (**238**)	*G. inflata*	root	(*305*)
	G. glabra	root	(*257*)
	G. eurycarpa	root	(*294*)
Glypallichalcone (**247**)	*G. pallidiflora*	root	(*316*)
Echinatin (**304**)	*G. glabra* (Xinjiang)	root	(*260*)
	G. aspera	root	(*293*)
	G. inflata	root	(*299*, *304*)
	G. pallidiflora	root	(*312*, *315*)
	G. squamulosa	root	(*456*)
	Northwest licorice		(*338*)
	Xinjiang licorice		(*287*)
	G. echinata (callus)		(*86*, *363*, *367*, *371*)
	G. uralensis (callus)		(*454*, *458*)
	G. glabra (callus, HPLC)		(*475*)
Kanzonol Y (**148**) (Dihydrochalcone)	*G. glabra* (Japanese)	root	(*252*)
	G. glabra (hairy root culture)		(*478*)
Gancaonin J (**244**) (Dihydrochalcone)	*G. pallidiflora*	root	(*312*)

Table 13 (*continued*)

Trivial name	Plant sources	Plant part(s)	References
4-Hydroxylonchocarpin (2′,4′-2 × OH, Dmp[3′,4′])	*G. glabra* (hairy root culture)		(*478*)
No name (**336**) (2′,4′-2 × OH, 4-O-Glu(2 ← 1)Api[5 ← *p*-coumaryl])	*G. uralensis*	root	(*479*)
No name (2′-OH, 4′-*O*-Glu, 4-*O*-Glu(2←1)-Api)	*G. aspera*	root	(*293*)
Dibenzoylmethane 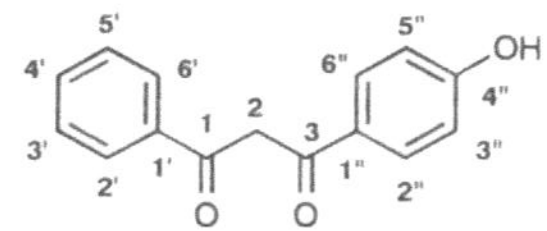			
Glyinflanin A (**216**)	*G. inflata*	root	(*295*)
(Glycyrdione A)	*G. inflata*	root	(*303*)
Glyinflanin B (**217**)	*G. inflata*	root	(*295*)
	G. eurycarpa	root	(*294*)
Glyinflanin C (**218**)	*G. inflata*	root	(*295*)
(Glycyrdione C)	*G. inflata*	root	(*302*)
Glyinflanin D (**219**)	*G. inflata*	root	(*295*)
Glyinflanin E (**220**)	*G. inflata*	root	(*299*)
Glyinflanin F (**221**)	*G. inflata*	root	(*299*)
Glycyrdione B (**228**)	*G. inflata*	root	(*303*)
Kanzonol A (**236**)	*G. eurycarpa*	root	(*294*)
5′-Prenyllicodione (**243**)	*G. inflata*	root	(*303*)
	G. eurycarpa	root	(*294*)
	G. echinata (callus)		(*86*, *366*)
2′-*O*-Methyllicodione (**246**)	*G. pallidiflora*	root	(*312*, *315*)
	synthesis		(*314*)
	synthesis (CMR)		(*365*)

Licodione (**305**)	*G. echinata* (callus)		(*364*, *367*)
	G. pallidiflora	root	(*315*)
	synthesis (CMR)		(*365*)
Flavanone	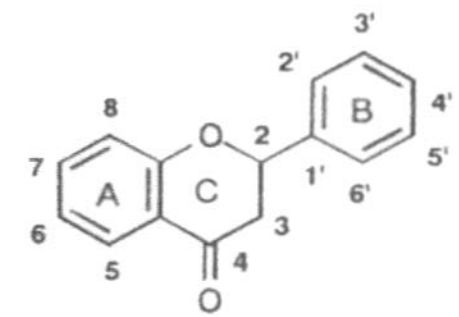		
Xambioona	*G. glabra*	root	(*257*)
(Dmp[3′,4′], Dmp[8,7])	*G. glabra* (hairy root culture)		(*478*)
Ovaliflavanone B	*G. pallidiflora*	root	(*312*)
(7-OH, 8-Prenyl)			
Euchrenone a_5	*G. inflata*	root	(*299*)
(7-OH, 8-Prenyl, Dmp[3′,4′])			
Pinocembrin (**328**)	*G. glabra*	n.r.	(*459*)
(5,7-2×OH)	*G. glabra*	leaf	(*341*, *460*)
	G. glabra	aerial part	(*461*)
	G. glabra (HPLC)	aerial part	(*462*)
	G. astragalina	aerial part	(*452*)
	G. lepidota	whole part	(*360*)
6-Prenylpinocembrin (**329**)	*G. glabra*	aerial part	(*461*)
(5,7-2×OH, 6-Prenyl)			
Pinostrobin (5-OH, 7-OMe)			
(Glyyunnanin)	*G. yunnanensis*	root	(*463*)
Sinoflavanone B	*G. pallidiflora*	root	(*312*)
(5,7-2×OH, 6,8-Diprenyl)			
Isobavachin	*G. pallidiflora*	root	(*78*, *312*)
(4′,7-2×OH, 8-Prenyl)			
No name	*G. pallidiflora*	root	(*78*, *312*, *315*, *455*)
(4′,7-2×OH, 6,8-Diprenyl)			

Trivial name	Plant sources	Plant part(s)	References
Naringenin (**325**)	*G. glabra*	aerial part	(*461*)
(4′,5,7-3×OH)	*G. eurycarpa*	aerial part	(*359*)
	Tiexin gancao		(*340*)
Choerospondin	*G. uralensis*	root	(*442a*)
(4′,5,7-3×OH, 4′-*O*-Glu)	*G. eurycarpa*	root	(*128*, *309*)
(Naringenin-4′-*O*-glucoside)			
Sophoraflavanone B	*G. uralensis*	aerial part	(*447*)
(4′,5,7-3×OH, 8-Prenyl)	*G. inflata*	root	(*211*)
	G. eurycarpa	aerial part	(*359*)
Citflavanone	*G. eurycarpa*	aerial part	(*359*)
(4′,5-2×OH, Dmp[8,7])			
Euchrestaflavanone A	*G. inflata*	root	(*299*)
(4′,5,7-3×OH, 3′,8-Prenyl)			
6-Prenylnaringenin	*G. glabra*	leaf	(*460*)
(4′,5,7-3×OH, 6-Prenyl)	*G. eurycarpa*	aerial part	(*359*)
Paratocarpin L	*G. uralensis*	aerial part	(*447*)
(4′,5,7-3×OH, 3′,6-Diprenyl)	*G. inflata*	root	(*299*)
Exiguaflavanone K	*G. eurycarpa*	aerial part	(*359*)
(4′,5,7-3×OH, 3′-OMe, 8-Prenyl)			
Sigmoidin B	*G. uralensis*	aerial part	(*346*)
(3′,4′5,7-4×OH, 5′-Prenyl)	*G. uralensis* (HPLC)	aerial part	(*462*)
	G. eurycarpa	aerial part	(*359*)
(Uralenin)	*G. uralensis*	leaf	(*353*)
Sigmoidin C	*G. eurycarpa*	aerial part	(*359*)
(3′,5,7-3×OH, Dmp[5′,4′])			
Liquiritin (**110**)	*G. glabra*	root	(*230*, *232*, *234*, *237*, *240*, *474*)
	G. glabra (HPLC)	root	(*228*, *239*)
	G. glabra (Xinjiang)	root	(*260*)
	G. glabra (Xinjiang, HPLC)	root	(*36*, *379*, *380*)
	G. uralensis	root	(*100*, *206*, *232a*, *238*)

	G. uralensis	n.r.	(*241*)
	G. uralensis (HPLC)	root	(*36*, *228*, *326*, *379*, *380*)
	G. aspera	root	(*98*)
	G. aspera (HPLC)	root	(*36*, *380*)
	G. inflata	root	(*210*, *211*, *433*, *448*)
	G. inflata	n.r.	(*451*)
	G. inflata (HPLC)	root	(*36*, *228*, *379*, *380*)
	G. eurycarpa	root	(*308*, *309*)
	G. eurycarpa (HPLC)	root	(*36*, *380*)
	G. echinata	root	(*180*)
	G. yunnanensis (HPLC)	root	(*36*)
	G. squamulosa (HPLC)	root	(*36*)
	G. pallidiflora (HPLC)	root	(*36*)
	Northwest licorice (HPLC)		(*326*, *329*)
Liquiritigenin (**111**)	*G. glabra*	root	(*232a*, *234*, *235*, *240*)
	G. glabra (HPLC)	root	(*239*)
	G. glabra (Xinjiang, HPLC)	root	(*36*, *379*)
	G. glabra	aerial part	(*342*)
	G. uralensis	root	(*232a*, *281*, *442a*, *464*)
	G. uralensis	n.r.	(*241*)
	G. uralensis (HPLC)	root	(*36*, *228*, *326*, *379*)
	G. aspera (HPLC)	root	(*36*)
	G. inflata	root	(*211*, *443*, *448*, *477*)
	G. inflata	n.r.	(*451*)
	G. inflata (HPLC)	root	(*36*, *379*)
	G. pallidiflora	root	(*312*, *315*)
	G. pallidiflora (HPLC)	root	(*36*)
	G. yunnanensis (HPLC)	root	(*36*)
	G. squamulosa (HPLC)	root	(*36*)
	G. eurycarpa (HPLC)	root	(*36*)
	G. echinata	root	(*180*)

Table 13 (*continued*)

Trivial name	Plant sources	Plant part(s)	References
	Northwest licorice (HPLC)		(*326*)
	G. uralensis (callus)		(*454*)
Neoliquiritin (**113**)	*G. glabra*	root	(*232a*, *465*)
	G. uralensis	root	(*232a*)
	G. uralensis	n.r.	(*241*)
	Northwest licorice (HPLC)		(*329*)
Liquiritin apioside (**115**)	*G. glabra*	root	(*237*)
	G. glabra (Xinjiang)	root	(*260*)
	G. glabra (HPLC)	root	(*239*)
	G. aspera (HPLC)	root	(*239*)
	G. uralensis	root	(*100*, *206*, *236*, *264*)
	G. inflata	root	(*128*, *210*)
	G. eurycarpa	root	(*128*, *309*)
(Apioliquiritin)	*G. uralensis* (HPLC)	root	(*326*)
Liquiritigenin 7-apiosylglucoside (**116**)	*G. glabra*	root	(*239*)
	G. aspera	root	(*293*)
Rhamno-liquiritin (**117**)	*G. glabra*	root	(*240*)
Glabrol (**127**)	*G. glabra*	root	(*250*, *253*, *254*)
	G. glabra (Xinjiang)	root	(*260*)
	G. glabra (Japanese)	root	(*252*)
	Xinjiang licorice	root	(*287*)
Glucoliquiritin apioside (**149**)	*G. glabra* (Xinjiang)	root	(*260*)
Shinflavanone (**152**)	*G. glabra* (Xinjiang)	root	(*260*)
Liquiritigenin 4′,7-diglucoside (**157**)	*G. uralensis*	root	(*100*, *264*, *266*)
6″-Acetylliquiritin (**158**)	*G. uralensis*	root	(*178*)
Glabranin (**267**)	*G. glabra*	aerial part	(*461*)
	G. glabra	n.r.	(*226*)
	G. lepidota	whole part	(*360*)
Licoflavanone (**268**)	*G. glabra*	leaf	(*341*, *460*)
	G. glabra (HPLC)	aerial part	(*462*)
Folerogenin (**269**)	*G. glabra*	aerial part	(*342*)

Gancaonin E (**278**)	*G. uralensis*	aerial part	(*346*)
Kanzonol S (**298**)	*G. eurycarpa*	aerial part	(*359*)
3-Hydroxyglabrol (**135**) (3-Hydroxyflavanone)	*G. glabra*	root	(*252*, *254*)
Glepidotin B (**303**) (3-Hydroxyflavanone)	*G. lepidota*	whole part	(*360*)
Eriodictyol (**316**)	*G. eurycarpa*	aerial part	(*359*)
No name (**309**)	*G. eurycarpa*	aerial part	(*359*)
(3′,4′,5,7-4×OH, 6-Prenyl)	synthesis		(*409*)
No name (**310**)	*G. uralensis*	leaf	(*356*)
(3′,4′,5,7-4×OH, 8-Prenyl)	*G. eurycarpa*	aerial part	(*359*)
	synthesis		(*409*)
No name (**337**) (7-OH, 8-Prenyl, 2HDmp[3′,4′]-3″-OH)	*G. glabra* (hairy root culture)		(*478*)
No name (**338**) (2*S*, 7-OH, 4′-O-Glu(2 ← 1)Api(5 ← *p*-coumaryl])	*G. uralensis*	root	(*479*)
No name (**339**) (2*R*, 7-OH, 4′-O-Glu(2 ← 1)Api(5 ← *p*-coumaryl])	*G. uralensis*	root	(*479*)
No name (**340**) (2*S*, 7-OH, 4′-O-Glu(2 ← 1)Api(5 ← indoleacetyl])	*G. uralensis*	root	(*479*)

Flavone

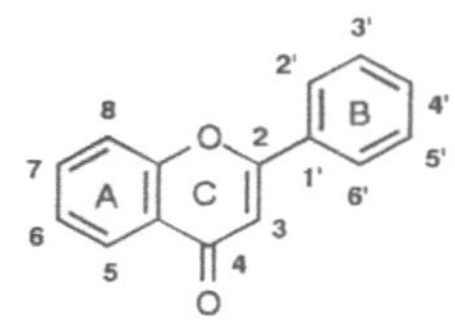

4′,7-Dihydroxyflavone	*G. uralensis*	root	(*447*)
	G. uralensis (HPLC)	root	(*380*)
	G. inflata	root	(*211*, *302*, *448*, *457*)
	G. inflata (HPLC)	root	(*380*)
	G. eurycarpa	root	(294)
	G. eurycarpa (HPLC)	root	(*380*)

Trivial name	Plant sources	Plant part(s)	References
	G. pallidiflora	root	(*315*, *455*)
	G. squamulosa	root	(*456*)
	Xinjiang licorice		(*287*, *335*)
	G. echinata (callus)		(*367*)
Sophoraflavone B (7-OH, 4′-*O*-Glu)	*G. eurycarpa*	root	(*309*)
No name (4′,7-2×OH, 8-Prenyl)	*G. echinata* (callus)		(*367*)
Genkwanin (4′,5-2×OH, 7-OMe)	*G. glabra*	aerial part	(*342*)
Vitexin (4′,5,7-3×OH, 8-*C*-Glu)	*G. echinata*	root	(*466*)
Isovitexin (4′,5,7-3×OH, 6-*C*-Glu)	*G. echinata*	root	(*342*)
Vicenin 2 (4′,5,7-3×OH, 6,8-*C*-diGlu)	*G. glabra* (HPLC)	aerial part	(*462*)
	G. uralensis	root	(*264*)
	G. uralensis	leaf	(*357*)
	G. eurycarpa	root	(*128*, *309*)
	G. pallidiflora (HPLC)	aerial parts	(*462*)
	G. echinata	root	(*466*)
Violanthin	*G. glabra* (Xinjiang, HPLC)	root	(*380*)
(4′,5,7-3×OH, 6-C-Glu, 8-C-Rha)	*G. uralensis* (HPLC)	root	(*380*)
	G. aspera (HPLC)	root	(*380*)
	G. inflata (HPLC)	root	(*380*)
	G. eurycarpa	root	(*128*, *309*)
	G. eurycarpa (HPLC)	root	(*380*)
Isoviolanthin	*G. glabra*	leaf, root	(*467*)
(4′,5,7-3×OH, 6-C-Rha, 8-C-Glu)	*G. uralensis* (HPLC)	root	(*380*)
	G. aspera (HPLC)	root	(*380*)
	G. inflata (HPLC)	root	(*380*)
	G. eurycarpa	root	(128, *309*)
	G. eurycarpa (HPLC)	root	(*380*)
	G. echinata	root	(*466*)
Schaftoside	*G. glabra*	leaf, root	(*467*)
(4′,5,7-3×OH, 8-C-Ara, 6-C-Glu)	*G. glabra* (Xinjiang, HPLC)	root	(*380*)
	G. glabra (HPLC)	aerial part	(*462*)

	G. uralensis (HPLC)	root	(*380*)
	G. aspera (HPLC)	root	(*380*)
	G. inflata (HPLC)	root	(*380*)
	G. eurycarpa	root	(*308*, *309*)
	G. eurycarpa (HPLC)	root	(*380*)
	G. echinata	root	(*466*)
	G. echinata (HPLC)	aerial part	(*462*)
	G. pallidiflora (HPLC)	aerial part	(*462*)
Isoschaftoside	*G. glabra*	leaf, root	(*467*)
(4′,5,7-3×OH,	*G. glabra* (Xinjiang, HPLC)	root	(*380*)
6-C-Ara, 8-C-Glu)	*G. uralensis* (HPLC)	root	(*380*)
	G. aspera (HPLC)	root	(*380*)
	G. inflata (HPLC)	root	(*380*)
	G. echinata	root	(*466*)
	G. eurycarpa (HPLC)	root	(*380*)
Prenyllicoflavone A (**151**)	*G. glabra* (Xinjiang)	root	(*260*)
	G. inflata	root	(*299*)
(Licoflavone B)	*G. inflata*	root	(*302*)
Licoflavone C (**227**)	*G. inflata*	root	(*211*, *302*, *477*)
No name (**234**)	*G. eurycarpa*	rhizome	(310)
Kanzonol D (**239**)	*G. eurycarpa*	root	(*294*)
Kanzonol E (**240**)	*G. eurycarpa*	root	(*294*)
Licoflavone A (**241**)	*G. glabra* (Xinjiang)	root	(*260*)
	G. uralensis (HPLC)	root	(*380*)
	G. inflata	root	(*302*, *448*)
	G. inflata (HPLC)	root	(*380*)
	G. eurycarpa	root	(*294*)
	Xinjiang licorice		(*74*)
	G. echinata (callus)		(*367*)
Gancaonin O (**282**)	*G. uralensis*	aerial part	(*347*)
Gancaonin Q (**284**)	*G. uralensis*	aerial part	(*348*)

Table 13 (*continued*)

Trivial name	Plant sources	Plant part(s)	References
Flavonol 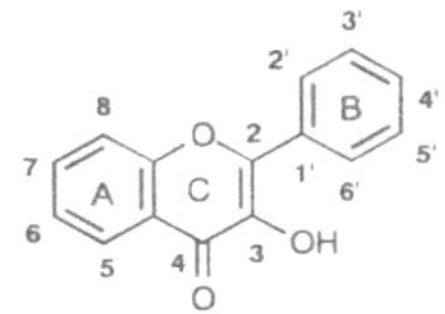			
Galangin (5,7-2×OH)	*G. glabra*	aerial part	(*461*)
Isokaemoferide (4′,5,7-3×OH, 3-OMe)	Northwest licorice		(*337*)
Kaempferol (4′,5,7-3×OH)	*G. glabra*	aerial part	(*342*)
	G. echinata	aerial part	(*468*)
Astragalin (Kaempferol 3-*O*-glucoside) (4′,5,7-3×OH, 3-*O*-Glu)	*G. glabra*	fruit, leaf	(*467*)
	G. glabra	aerial part	(*342*)
	G. glabra (HPLC)	aerial part	(*462*)
	G. uralensis	leaf	(*357*)
	G. uralensis (HPLC)	aerial part	(*462*)
	G. pallidiflora (HPLC)	aerial part	(*462*)
	G. echinata	aerial part	(*468*)
Nicotiflorin (Kaempferol-3-*O*-rutinoside) (4′,5,7-3×OH, 3-*O*-Rutinosyl)	*G. uralensis*	leaf	(*357*)
	G. echinata	aerial part	(*468*)
	G. echinata (HPLC)	aerial part	(*462*)
Kaempferol-3-*O*-rhamnosylgalactoside	*G. echinata* (HPLC)	aerial part	(*462*)
Isovitexin (Saponaretin) (4′,5,7-3×OH, 6-C-Glu)	*G. glabra*	leaf	(*342*)
Isorhamnetin (4′,5,7-3×OH, 3′-OMe)	*G. glabra*	aerial part	(*342*)

Fisetin (3′,4′,7-3×OH)	*G. uralensis*	root	(*286*)
Quercetin (3′,4′,5,7-4×OH)	*G. glabra*	aerial part	(*342*)
	G. uralensis	leaf	(*354*, *355*)
	G. echinata	aerial part	(*468*)
No name (3′,4′5,7-4×OH, 3-OMe, 6-Prenyl)	*G. eurycarpa*	aerial part	(*359*)
Quercetin 3,4′-dimethyl ether (3′,5,7-3 × OH, 3,4′-2 × OMe)	*G. eurycarpa*	aerial part	(*359*)
Quercetin 3,3′-dimethyl ether (4′,5,7-3×OH, 3,3′-2×OMe)	*G. uralensis*	leaf	(*353*)
	G. eurycarpa	aerial part	(*359*)
Isoquercitrin (3′,4′,5,7-4×OH, 3-*O*-Glu)	*G. glabra*	leaf	(*344*, *467*)
	G. glabra	aerial part	(*342*)
	G. glabra	fruit	(*467*)
	G. glabra (HPLC)	aerial part	(*462*)
	G. uralensis	leaf	(*357*)
	G. uralensis (HPLC)	aerial part	(*462*)
	G. pallidiflora (HPLC)	aerial part	(*462*)
	G. echinata	aerial part	(*468*)
	G. echinata (HPLC)	aerial part	(*462*)
Quercetin-3-*O*-glucobioside	*G. glabra*	aerial part	(*342*)
Cacticin (3′-*O*-Glu, 4′,5,7-3×OH, 3′-O Me)	*G. pallidiflora* (HPLC)	aerial part	(*462*)
Narcissin (4′,5,7-3×OH, 3′-OMe, 3-*O*-Rutinosyl) (Isorhamnetin-3-*O*-rutinoside)	*G. uralensis*	leaf	(*357*)
	G. echinata (HPLC)	aerial part	(*462*)
Rutin (Quercetin-3-*O*-rutinoside)	*G. glabra* (HPLC)	aerial part	(*462*)
	G. uralensis	leaf	(*357*)
	G. uralensis (HPLC)	aerial part	(*462*)
	G. echinata	aerial part	(*468*)
	G. echinata (HPLC)	aerial part	(*462*)

Table 13 (*continued*)

Trivial name	Plant sources	Plant part(s)	References
Isolicoflavonol (**171**)	*G. uralensis*	root	(*281*)
	Northwest licorice		(*335*)
Glyasperin A (**193**)	*G. aspera*	root	(*288*)
Topazolin (**207**)	*G. aspera*	root	(*288*)
	Northwest licorice		(*336*)
Licoflavonol (**208**)	*G. uralensis*	root	(*282*)
	G. aspera	root	(*291*)
	Northwest licorice		(*333*)
	Northwest licorice (HPLC)		(*329*)
	synthesis		(*276*)
Kumatakenin (**262**)	*G. uralensis*	root	(*282*)
	G. aspera	root	(*288*)
	Northwest licorice (HPLC)		(*329*)
	Northwest licorice		(*277*, *333*)
Glyphoside (**270**)	*G. glabra*	aerial part	(*342*)
	G. glabra	leaf	(*344*)
Kaempferol-3-*O*-diglucoside (**271**)	*G. glabra*	aerial part	(*342*)
Astragalin monoacetate (**272**)	*G. glabra*	aerial part	(*342*)
Gancaonin P (**283**)	*G. uralensis*	aerial part	(*347*)
	synthesis		(*276*)
Ganocanin P 3′-*O*-methyl ether (**283a**)	*G. uralensis*	leaf	(*356*)
Uralenol (**290**)	*G. uralensis*	leaf	(*353*)
Uralenol 3-*O*-methyl ether (**291**)	*G. uralensis*	leaf	(*354*, *355*)
Neouralenol (**292**)	*G. uralensis*	leaf	(*353*)
Uralene (**293**)	*G. uralensis*	leaf	(*354*, *355*)
Glepidotin A (**301**)	*G. lepidota*	whole part	(*360*)
	synthesis		(*276*)

Isoflavanone	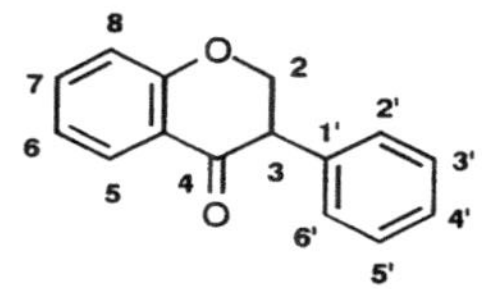		
3′,8-Diprenyldalbergioidin	*G. aspera*	root	(*288*)
(2′,4′,5,7-4×OH,	*G. glabra* (Xinjiang)	root	(*262*)
3′,8-Diprenyl)	*G. uralensis* (Xinjiang)	root	(ud)
	Northwest licorice		(*336*)
RL-Q (**138**)	*G. glabra*	root	(*259*)
RL-R (**139**)	*G. glabra*	root	(*259*)
Kanzonol G (**179**)	*G. uralensis*	root	(*284*)
Licoisoflavanone (**192**)	*G. glabra* (Xinjiang)	root	(*263*)
	G. aspera	root	(*288*)
	Xinjiang licorice		(*287*)
Glyasperin B (**194**)	*G. aspera*	root	(*288*)
Glyasperin F (**198**)	*G. aspera*	root	(*290*)
Glyasperin J (**202**)	*G. aspera*	root	(*290*)
Glyasperin K (**203**)	*G. uralensis*	root	(ud)
	G. aspera	root	(*291*)
Glyasperin M (**205**)	*G. glabra* (Xinjiang)	root	(*263*)
	G. aspera	root	(*291*)
Glycyrrhisoflavanone (**253**)	Northwest licorice		(*335*)
No name (**254**)	*G. glabra* (Xinjiang)	root	(*262*)
	G. uralensis (Xinjiang)	root	(ud)
	Northwest licorice		(*336*)

Table 13 (*continued*)

Trivial name	Plant sources	Plant part(s)	References
Isoflavone	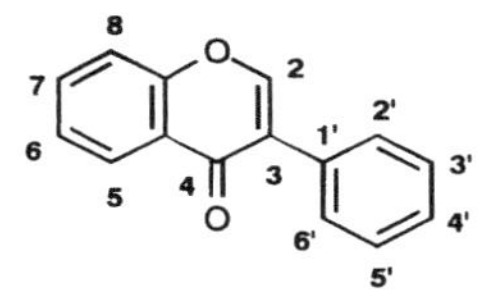		
4′,7-di-*O*-Methyldaidzein (4′,7-2×OMe)	*G. pallidiflora*	root	(*312*)
Ononin (4′-OMe, 7-*O*-Glu)	*G. glabra* (HPLC)	aerial part	(*462*)
	G. glabra (Xinjiang, HPLC)	root	(*380*)
	G. glabra (Xinjiang)	root	(*260*)
	G. uralensis	root	(*206, 238*)
	G. uralensis (HPLC)	root	(*326, 380*)
	G. uralensis (HPLC)	aerial part	(*462*)
	G. aspera (HPLC)	root	(*380*)
	G. inflata	root	(*448*)
	G. inflata (HPLC)	root	(*380*)
	G. eurycarpa	root	(*128, 309*)
	G. eurycarpa (HPLC)	root	(*380*)
	G. pallidiflora	root	(*315*)
	G. pallidiflora	aerial part	(*447*)
	Northwest licorice (HPLC)		(*326*)
	G. echinata (callus)		(*86*)
Afromosin (7-OH, 4′,6-2×OMe)	*G. pallidiflora*	root	(*78, 312, 315*)
	G. squamulosa	root	(*456*)
Calycosin (3′,7-2×OH, 4′-OMe)	*G. pallidiflora*	root	(*312, 455*)
	G. uralensis (callus)		(*454*)

Isoderrone	*G. inflata*	root	*(299)*
(5,7-2×OH, Dmp[3′, 4′])	*G. uralensis* (Xinjiang)	root	(ud)
8-*O*-Methylretusin	Tiexin gancao		*(340)*
(7-OH, 4′,8-2×OMe)			
Prunetin	*G. glabra*	n.r.	*(459)*
(4′,5-2×OH, 7-OMe)	*G. glabra*	leaf	*(460)*
	G. pallidiflora	aerial part	*(312)*
No name (2′,7-2×OH, 4′-OMe)	*G. pallidiflora*	root	*(315)*
Genistein (4′,5,7-3×OH)	*G. glabra*	aerial part	*(461)*
	G. glabra	leaf	*(460)*
	G. glabra (HPLC)	aerial part	*(462)*
	Northwest licorice		*(338)*
Wighteone	*G. glabra*	leaf	*(460)*
(4′,5,7-3×OH, 6-Prenyl)	*G. glabra* (HPLC)	aerial part	*(462)*
Lupiwighteone	*G. glabra*	leaf	*(460)*
(4′,5,7-3×OH, 8-Prenyl)	*G. glabra* (HPLC)	aerial part	*(462)*
	G. uralensis	aerial part	*(346)*
Hydroxywighteone	*G. glabra* (HPLC)	aerial part	*(462)*
(4′,5,7-3×OH,	*G. uralensis*	aerial part	*(447)*
6-Prenyl(5-OH) [trans])			
3′-Dimethylallylgenistein	*G. aspera*	root	*(291)*
(4′,5,7-3×OH, 3′-Prenyl)			
Sophoraisoflavone A	*G. inflata*	root	*(447)*
(4′,5,7-3×OH, Dmp[3′,2′])			
8-(γ, γ-dimethylallyl)-Wighteone	*G. inflata*	root	*(299)*
(4′,5,7-3×OH, 6,8,-Diprenyl)	*G. glabra* (Xinjiang)	root	*(263)*
	G. uralensis (Xinjiang)	root	(ud)
	Northwest licorice		*(447)*
Lupalbigenin	*G. glabra* (Xinjiang)	root	*(263)*
(4′,5,7-3×OH, 3′,6-Diprenyl)			
2′,7-Dihydroxy-4′,6-dimethoxyisoflavone	*G. pallidiflora*	root	*(312)*

Table 13 (*continued*)

Trivial name	Plant sources	Plant part(s)	References
Angustone B (2′5,7-3×OH, 6-Prenyl, Dmp[3′,4′])	*G. inflata*	root	(*299*)
Wistin (4′,6-2×OMe, 7-O-Glu)	*G. pallidiflora*	root	(*315*)
7-Hydroxy-2-methylisoflavone (**121**)	*G. glabra* (India)	root	(*246*)
7-Acetoxy-2-methylisoflavone (**121a**)	*G. glabra* (India)	root	(*246*)
7-Methoxy-2-methylisoflavone (**122**)	*G. glabra* (India)	root	(*246*)
Glyzarin (**123**)	*G. glabra* (India)	root	(*249*)
Glyzaglabrin (**125**)	*G. glabra* (India)	root	(*248*)
Glabrone (**128**)	*G. glabra*	root	(*251*)
	G. inflata	root	(*211*, *295*, *477*)
Kanzonol T (**156**)	*G. glabra* (Xinjiang)	root	(*263*)
Isoononin (**160**)	*G. uralensis*	root	(*266*)
	G. inflata	root	(*211*, *457*)
Licoricone (**162**)	*G. uralensis*	root	(*269*, *279*, *280*, *464*)
	Northwest licorice		(*277*, *333*, *334*)
	X-ray		(*270*)
	synthesis		(*271*)
Semilicoisoflavone B (**173**)	*G. uralensis*	root	(*282*, *286*)
	G. aspera	root	(*288*)
Isoangustone A (**174**)	*G. uralensis*	root	(*282*)
Kanzonol K (**183**)	*G. uralensis*	root	(*285*)
Kanzonol L (**184**)	*G. uralensis*	root	(*285*)
Glyasperin N (**206**)	*G. aspera*	root	(*291*)
Licoisoflavone B (**215**)	*G. glabra* (Xinjiang)	root	(*263*)
	G. uralensis (Xinjiang)	root	(ud)
	G. aspera	root	(*288*)
	G. inflata	root	(*299*)
	Xinjiang licorice		(*287*)
Glycyroside (**233**)	*G. eurycarpa*	root	(*308*, *309*)
Eurycarpin A (**235**)	*G. eurycarpa*	root	(*311*)

References, pp. 110–140

Formononetin (**242**)	*G. glabra*	root	(*251*, *253*, *254*, *469*)
	G. uralensis	root	(*280*, *330*)
	G. uralensis	aerial part	(*346*)
	G. inflata	root	(*211*, *477*)
	G. eurycarpa	root	(ud)
	G. pallidiflora	root	(*78*, *312*, *315*, *316*, *455*)
	G. pallidiflora	aerial part	(*312*)
	G. squamulosa	root	(*456*)
	G. yunnanensis	n.r.	(453)
	Northwest licorice		(*447*)
	Xinjiang licorice		(*287*)
	Tiexin gancao		(*340*)
	G. glabra (callus)		(*131*, *470*)
	G. uralensis (callus)		(*454*)
	G. echinata (callus)		(*86*, *369*)
	G. glabra (callus, HPLC)		(*475*)
	licorice		(*473*)
Licoisoflavone A (**249**)	*G. aspera*	root	(*291*)
	Northwest licorice		(*334*)
Glycyrrhisoflavone (**250**)	Northwest licorice		(*335*)
Glisoflavone (**251**)	Northwest licorice		(*337*)
Glicoricone (**252**)	Northwest licorice		(*338*)
Gancaonin G (**259**)	*G. glabra* (Xinjiang)	root	(*263*)
	G. uralensis (Xinjiang)	root	(ud)
	Northwest licorice		(*277*)
Gancaonin H (**260**)	*G. glabra* (Xinjiang)	root	(*263*)
	Northwest licorice		(*277*)
7-*O*-Methylluteone (**263**)	*G. uralensis* (Xinjiang)	root	(ud)
Gancaonin A (**274**)	*G. uralensis*	aerial part	(*346*)
Gancaonin B (**275**)	*G. uralensis*	aerial part	(346)
Gancaonin C (**276**)	*G. uralensis*	aerial part	(*346*)

Table 13 (*continued*)

Trivial name	Plant sources	Plant part(s)	References
Gancaonin D (**277**)	*G. uralensis*	aerial part	(*346*)
Gancaonin L (**279**)	*G. uralensis*	aerial part	(*347*)
Gancaonin M (**280**)	*G. uralensis*	aerial part	(*347*)
Gancaonin N (**281**)	*G. uralensis*	aerial part	(*347*)
Pallidiflorin (5-OH, 4′-OMe)	*G. pallidiflora*	root	(*196*)
Pterocarpan	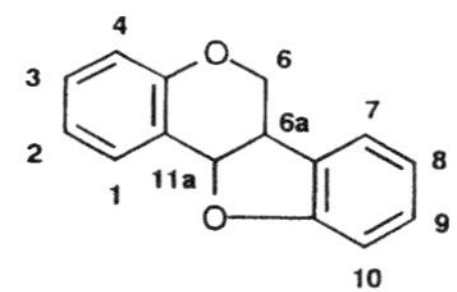		
Medicarpin (3-OH, 9-OMe)	*G. galbra* (Xinjiang)	root	(*260*)
	G. uralensis	root	(*282*)
	G. pallidiflora	root	(*78*, *312*, *315*, *455*)
	G. yunnanensis	n.r.	(*453*)
	Tiexin gancao		(*340*)
Medicarpin 3-*O*-glucoside (3-*O*-Glu, 9-OMe)	*G. eurycarpa*	root	(*309*)
Homopterocarpin (3,9-2×OMe)	*G. pallidiflora*	root	(*198*, *199*)
	G. yunnanensis	root	(*463*)
	G. yunnanensis	n.r.	(*453*)
Phaseollin (3-OH, Dmp[10, 9])	*G. inflata*	root	(*299*)
1-Methoxyphaseollidin (1-OMe, 3,9-2×OH, 10-Prenyl)	*G. uralensis*	root	(*282*)
ent-(–)-Hemileiocarpin (9-OMe, Dmp[4,3])	*G. glabra* (Xinjiang)	root	(*260*)

Shinpterocarpin (**153**)	*G. glabra* (Xinjiang)	root	(*260*)
	G. glabra (Japanese)	root	(*252*)
1-Methoxyphaseollin (**154**)	*G. glabra* (Xinjiang)	root	(*260*)
1-Methoxyficifolinol (**176**)	*G. uralensis*	root	(*282*, *284*)
	G. aspera	root	(*288*, *293*)
	Tiexin gancao		(*340*)
Kanzonol F (**178**)	*G. uralensis*	root	(*284*)
	Tiexin gancao		(*340*)
Kanzonol P (**188**)	*G. uralensis*	root	(*261*)
No name (**212**)	*G. aspera*	root	(*293*)
No name (**341**) (3-OH, 9-OMe, 4-Prenyl)	*G. glabra* (hairy root culture)		(*478*)

Coumestan

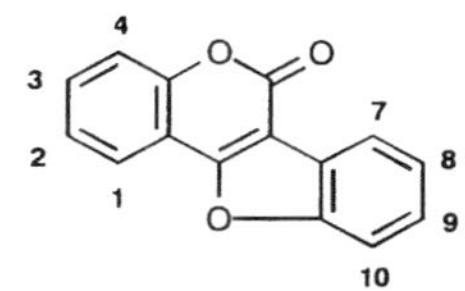

4′-*O*-Methylcoumesterol (3-OH, 9-OMe)	*G. pallidiflora*	root	(*316*)
3-*O*-Methylglycyrol (**161**)	*G. uralensis*	root	(*278*, *282*)
	Northwest licorice		(277, *334*)
	licorice		(*268*)
Isoglycyrol (**168**)	*G. uralensis*	root	(*278*, *281*, *282*, *330*)
	G. uralensis (HPLC)	root	(*36*, *327*, *379*)
	G. glabra (Xinjiang, HPLC)	root	(*379*)
	G. aspera	root	(*98*)
	G. aspera (HPLC)	root	(*36*)
	G. inflata (HPLC)	root	(*36*, *379*)
	licorice		(*268*)

Table 13 (*continued*)

Trivial name	Plant sources	Plant part(s)	References
	Northwest licorice		(*277*)
	Northwest licorice (HPLC)		(*327*)
Glycyrol (**169**)	*G. glabra* (Xinjiang, HPLC)	root	(*36*, *379*)
	G. uralensis	root	(*281*, *278*, *330*)
	G. uralensis (HPLC)	root	(*36*, *327*, *379*)
	G. aspera	root	(*98*, *291*)
	G. aspera (HPLC)	root	(*36*)
	G. yunnanensis (HPLC)	root	(*36*)
	G. squamulosa (HPLC)	root	(*36*)
	Northwest licorice		(*277*, *327*, *334*)
	Northwest licorice (HPLC)		(*329*)
	licorice		(*268*)
	Tiexin gancao		(*340*)
(Neoglycyrol)	*G. uralensis*	root	(*464*)
Gancaonin F (**258**)	Northwest licorice		(*277*)
Isoflavan	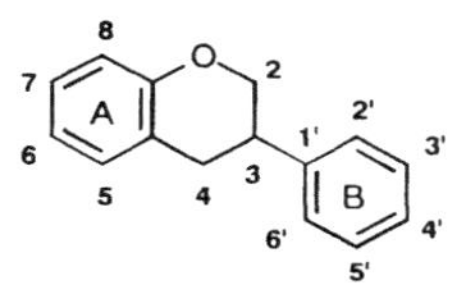		
Vestitol	*G. echinata*	root	(*78*)
(2′,7-2×OH, 4′-OMe)	*G. pallidiflora*	root	(*78*, *315*)
	Northwest licorice		(78)
Phaseollinisoflavan	*G. glabra*	root	(*253*, *254*)
(2′,7-2×OH, Dmp[3′,4′])			
Glabridin (**126**)	*G. glabra*	root	(*250*, *253*, *254*, *255*, *434*)
	G. glabra (Japanese)	root	(*252*)

	G. glabra (HPLC)	root	(*228, 471*)
	G. uralensis	root	(*443*)
	G. glabra (callus, HPLC)		(*475*)
Hispaglabridin A (**131**)	*G. glabra*	root	(*253, 254, 255*)
	G. glabra (Xinjiang)	root	(*260*)
	G. glabra (Japanese)	root	(*252*)
Hispaglabridin B (**132**)	*G. glabra*	root	(*253, 254, 255*)
	G. glabra (Xinjiang)	root	(*260*)
4′-*O*-Methylglabridin (**133**)	*G. glabra*	root	(*253, 254, 255*)
	G. glabra (Japanese)	root	(*252*)
	G. glabra (hairy root culture)		(*478*)
8-Prenyl-phaseollinisoflavan (**136**)	*G. glabra*	root	(*255*)
3′-Hydroxy-4′-*O*-methylglabridin (**143**)	*G. glabra*	root	(*254, 255*)
Kanzonol X (**147**)	*G. glabra* (Japanese)	root	(*252*)
(Tenulfolin B)			
Methylhispaglabridin B (**150**)	*G. glabra* (Xinjiang)	root	(*260*)
Kanzonol R (**155**)	*G. glabra* (Xinjiang)	root	(*261*)
Licoricidin (**164**)	*G. uralensis*	root	(*279, 280, 282, 284, 443*)
	G. aspera	root	(*288*)
	Northwest licorice		(*272, 334*)
	Tiexin gancao		(*340*)
	licorice		(*267*)
(Licorisoflavan B)	licorice		(*274*)
Licorisoflavan A (**165**)	*G. uralensis*	root	(*273, 282, 284, 440*)
	G. aspera	root	(*288*)
	Tiexin gancao		(*340*)
	licorice		(*274*)
	synthesis		(*275*)
Kanzonol H (**180**)	*G. uralensis*	root	(*284*)
Kanzonol I (**181**)	*G. uralensis*	root	(*284*)
	Tiexin gancao		(*340*)
Kanzonol J (**182**)	*G. uralensis*	root	(*284*)
Kanzonol M (**185**)	*G. uralensis*	root	(*261*)

Table 13 (*continued*)

Trivial name	Plant sources	Plant part(s)	References
Kanzonol N (**186**)	*G. uralensis*	root	(*261*)
Kanzonol O (**187**)	*G. uralensis*	root	(*261*)
Glyasperin C (**195**)	*G. aspera*	root	(*288*, *293*)
	G. uralensis (Xinjiang)	root	(ud)
Glyasperin D (**196**)	*G. uralensis*	root	(*284*)
	G. aspera	root	(*288*)
	synthesis		(*275*)
	Northwest licorice		(*336*)
Glyasperin G (**199**)	*G. aspera*	root	(*290*)
Glyasperin H (**200**)	*G. aspera*	root	(*290*)
Glyasperin I (**201**)	*G. aspera*	root	(*290*)
Glyinflanin I (**224**)	*G. inflata*	root	(*299*)
Glyinflanin J (**225**)	*G. inflata*	root	(*299*)
Glyinflanin K (**226**)	*G. inflata*	root	(*299*)
	G. glabra (hairy root culture)		(*478*)
Gancaonin X (**264**)	Tiexin gancao		(*340*)
Gancaonin Y (**265**)	Tiexin gancao		(*340*)
Gancaonin Z (**266**)	*G. inflata*	root	(*299*)
	Tiexin gancao		(*340*)
Isomucronulatol (**273**)	*G. glabra* (leaf, phytoalexin)		(*345*, *472*)
No name (**342**) (2′ OH, 4′-OMe, 7-O-Glu)	*G. uralensis*	root	(*479*)
Isoflav-3-ene	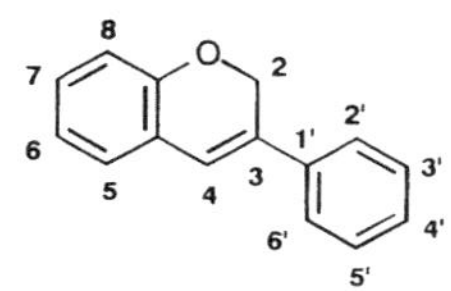		
Pallidiflorene (2′,7-2×OH, 4′-OMe)	*G. pallidiflora*	root	(*78*, *315*)

Glabrene (**130**)	*G. glabra*	root	(*251, 253, 254, 434*)
	G. glabra (HPLC)	root	(*228*)
	G. glabra (Japanese)	root	(*252*)
	G. glabra (Xinjiang)	root	(*260*)
RL-S (**140**)	*G. glabra*	root	(*259*)
Dehydroglyasperin C (**211**)	*G. aspera*	root	(*293*)
2-Arylbenzofuran	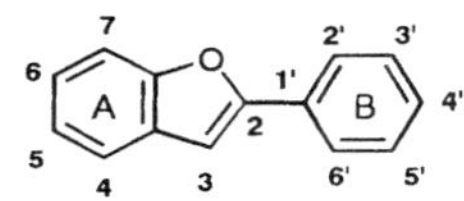		
Kanzonol U (**144**)	*G. glabra* (Japanese)	root	(*252*)
(Glabrocoumarone A)	*G. glabra*	root	(*257*)
Kanzonol V (**145**)	*G. glabra* (Japanese)	root	(*252*)
Liconeolignan (**170**)	*G. uralensis*	root	(*280*)
(Licobenzofuran)	*G. uralensis*	root	(*279*)
Gancaonin I (**177**)[b]	*G. uralensis*	root	(*282*)
	G. aspera	root	(*447*)
	Northwest licorice		(*277*)
Kanzonol I (**181**)	*G. uralensis*	root	(*284*)
	Tiexin gancao	.	(*340*)
Licocoumarone (**290**)	*G. uralensis* (HPLC)	root	(*228*)
	G. aspera	root	(*288, 293*)
	Northwest licorice		(*332, 337*)
Glyinflanin H (**223**)	*G. inflata*	root	(*299*)
(Glabrocoumarone B)	*G. glabra*	root	(*257*)
Licofuranone (**257**) (2-Arylbenzofuranone)	Northwest licorice		(*338*)

Table 13 (*continued*)

Trivial name	Plant sources	Plant part(s)	References
3-Arylcoumarin	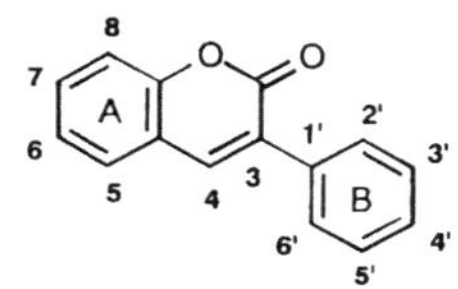		
2′,7-Dihydroxy-4′-methoxy-3-arylcoumarin	*G. pallidiflora*	root	(*315*)
RL-P (**137**)	*G. glabra*	root	(*259*)
RL-U (**141**)	*G. glabra*	root	(*259*)
Kanzonol W (**146**)	*G. glabra* (Japanese)	root	(*252*)
	G. glabra	root	(*259*)
Glycycoumarin (**172**)	*G. glabra* (Xinjiang, HPLC)	root	(*36*, *379*)
	G. uralensis	root	(*281*, *330*)
	G. uralensis (HPLC)	root	(*36*, *228*, *327*, *379*)
	G. aspera	root	(*98*, *291*, *293*)
	G. aspera (HPLC)	root	(*36*)
	G. inflata (HPLC)	root	(*36*, *379*)
	G. eurycarpa (HPLC)	root	(*36*)
	G. yunnanensis (HPLC)	root	(*36*)
	G. squamulosa (HPLC)	root	(*36*)
	Northwest licorice		(*327*, *332*, *335*)
No name (**191**)	*G. uralensis*	root	(*286*)
Glyasperin L (**204**)	*G. aspera*	root	(*291*)
Isoglycycoumarin (**210**)	*G. glabra* (Xinjiang, HPLC)	root	(*36*, *379*)
	G. uralensis (HPLC)	root	(*36*, *379*)
	G. aspera	root	(*98*)
	G. aspera (HPLC)	root	(*36*)
	G. inflata (HPLC)	root	(*36*, *379*)
	G. eurycarpa (HPLC)	root	(*36*)
	G. yunnanensis (HPLC)	root	(*36*)

Glycyrin (**248**)	*G. uralensis*	root	(*330, 443*)
	Northwest licorice		(*283*)
	Northwest licorice (HPLC)		(*327*)
Licoarylcoumarin (**255**)	*G. uralensis*	root	(*443*)
	Northwest licorice		(*337*)
Licopyranocoumarin (**256**)	*G. uralensis*	root	(*442a*)
	G. uralensis (HPLC)	root	(*228*)
	licorice	root	(*439*)
	Northwest licorice		(*337, 427*)
Gancaonin W (**261**)	*G. aspera*	root	(*291*)
	Northwest licorice		(*339*)

Dihydrostilbene

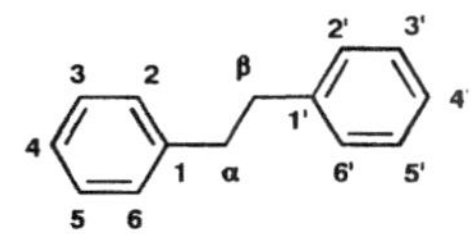

3,5-Dihydroxy-4-(3-methyl-2-butenyl)-bibenzyl (3,5-2×OH, 4-Prenyl)	*G. lepidota*	whole part	(*360*)
Gancaonin R (**285**)	*G. uralensis*	aerial part	(*348*)
Gancaonin S (**286**)	*G. uralensis*	aerial part	(*348*)
Gancaonin T (**287**)	*G. uralensis*	aerial part	(*348*)
No name (**294**)	*G. acanthocarpa*	aerial part	(*358*)
No name (**295**)	*G. acanthocarpa*	aerial part	(*358*)
Glepidotin C (**299**)	(*G. lepidota*)	whole part	(*361*)

Table 13 (*continued*)

Trivial name	Plant sources	Plant part(s)	References
Dihydrophenanthrene	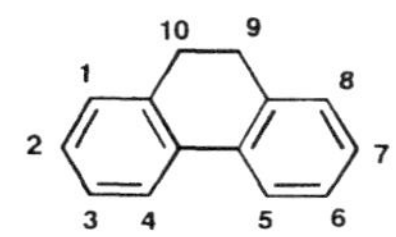		
Gancaonin U (**288**)	*G. uralensis*	aerial part	(*348*)
Gancaonin V (**289**)	*G. uralensis*	aerial part	(*348*)
Others			
5-(1,1-Dimethylallyl)-4-hydroxy-2-methoxybenzaldehyde	*G. inflata*	root	(ud)
5,7-Dihydroxy-2-methylchromone	*G. pallidiflora*	aerial part	(*447*)
5,7-Dihydroxy-2-methylchromone 7-*O*-glucoside	*G. pallidiflora*	aerial part	(*447*)
Umbelliferone (Coumarin, 7-OH)	*G. glabra* (HPLC)	n.r.	(*476*)
Scopoletin (Coumarin, 6-OMe, 7-OH)	*G. uralensis*	aerial part	(*346*)
Liqcoumarin (**124**) (Coumarin)	*G. glabra* (India)	root	(*247*)
RL-V (**142**)	*G. glabra*	root	(*259*)
Licoriphenone (**175**)	*G. uralensis*	root	(*282*)
Kanzonol Q (**189**) (Coumarin)	*G. uralensis*	root	(*261*)
Glyasperin E (**197**)	*G. aspera*	root	(*289*)
Inflacoumarin A (**232**)[c] (Neoflavone)	*G. inflata*	root	(*211*, *306*, *307*)
Gancaonin K (**245**)	*G. pallidiflora*	root	(*312*)
No name (**296**)	*G. acanthocarpa*	aerial part	(*358*)

No name (**297**)	*G. acanthocarpa*	aerial part	(*358*)
Uralenneoside (Benzene, 3,4-2×OH, 1-COO-Xyl)	*G. uralensis*	leaf	(*357*)
No name (**343**) (Bibenzoyl, 2,2′,4-3 × OH, 4′-OMe, 3-Prenyl)	*G. glabra* (hairy root culture)		(*478*)
No name (**344**) (Bisflavone; aurone (3′,4′,6-3 × OH, 7-Prenyl)[6′-2]2-benzoylcoumaranone(4′,6-2 × OH, 3,3′-Diprenyl)	*G. glabra* (hairy root culture)		(*478*)
Carpusin (Auronol, 4′,6-2 × OH, 4-OMe)	*G. uralensis*	root	(*479*)

[a] The trivial name "licuraside" is used in Chemical Abstracts and "liquiraside" is used in Doklady (Engl. transl.). The authors used the different numbering system (B ring: C1′–C6′); [b] The ^{1}H NMR spectrum of the compound (**177**) is similar to that of liconeolignan (**170**), but the IR spectra of these compounds are different each other (unpublished data); [c] The spectra data (UV, MS, ^{1}H and ^{13}C NMR spectra) of the compound (**232**) are similar to those of licoflavone A (**241**). The difference between **232** and **241** may be shown only with their HMBC spectra (*307*)

Acknowledgments

The authors thanks Professor WERNER HERZ (The Florida State University) who patiently revised the text and gave useful advice. Thanks are also due to Dr. SHOJI SHIBATA (Shibata Laboratory of Natural Medicinal Materials) for critically reading this manuscript and for helpful suggestions.

Note added in proof

This review article covers constituents of licorice (*Glycyrrhiza* species) reported up to the end of 1996. After submitting the review, the isolations of nine new phenols (**336–344**) from hairy root cultures of *G. glabra* or roots of *G. uralensis* were reported [*478*, *479*]. The structures are added in Table 13 along with known compounds in the reports.

References

1. HARBORNE, J.B.: Phytochemistry of the Leguminosae, In: Phytochemical Dictionary of the Leguminosae, vol. 1. Plants and their Constituents. (BISBY, F.A., J. BUCKINGHAM, and J.B. HARBORNE, eds.), pp. xx–xxiii. London: Chapman and Hall, 1994.
2. POLHILL, R.M.: Classification of the Leguminosae. In: Phytochemical Dictionary of the Leguminosae, vol. 1. Plants and their Constituents (BISBY, F.A., J. BUCKINGHAM, and J.B. HARBORNE, eds.), pp. xxxv–xlii. London: Chapman and Hall, 1994.
3. SCHOFIELD, M.: The History of *Glycyrrhiza*. Pharm. J., **165**, 76 (1950).
4. GIBSON, M.R.: *Glycyrrhiza* in Old and New Perspectives. J. Nat. Prod. (Lloydia), **41**, 348 (1978).
5. FENWICK, G.R., J. LUTOMSKI, and C. NIEMAN: Liquorice, *Glycyrrhiza glabra* L. Composition, Uses and Analysis. Food Chem., **38**, 119 (1990).
6. LUTOMSKI, J., C. NIEMAN, and G.R. FENWICK: Liquorice, *Glycyrrhiza glabra* L. Biological Properties. Herba Polonica, **37**, 163 (1991).
7. TAKAHASHI, S.: Kanpo-yaku to Sono Hattensi (Chinese Medicine and History of its Development), Toyama (Japan): Kougensha, 1976.
8. KUBO, M., and T. TANI: Kanpoh Iyaku-gaku (Traditional Sino-Medication), pp. 1–25, Tokyo: Hirokawa-Shoten, 1985.
9. OTSUKA, Y.: Introductory Remarks on Licorice from the Viewpoint of Medical History. Taisha (Metabolism Disease, Japan), **10**, 613 (1973).
10. OOTUKA, K.: Rinshou Ouyou Shokan-Ron Kaisetu (Clinical Application and Annotation of Shang Han Lun). Tokyo: Sogensha, 1966.
11. MAEDA, T.: Licorice. Koryo (Flavor, Japan), **134**, 53 (1981).
12. VORA, P.S.: Characteristics and Applications of Licorice Products in Tobacco. Tab. Int., **186**, 13 (1984).
13. TATUSAKI, T.: Licorice in Phytochemical Industry. In: Records of Lecture in 2nd Meeting on Crude Drugs; Licorice, p. 58. Tokyo: Tokyo Soc. Pharmacognosy, 1986.
14. KUSANO, G., and M. SHIBANO: A Try to Revive *Glycyrrhiza uralensis* in Kanzoh-yashiki. Foods Food Ingredients J. Jpn., **161**, 73 (1994).
15. KURAMOTO, T., and M. YAMAMOTO: Monoglucuronylglycyrrhetic Acid (MGGR) from Glycyrrhizic Acid by β-Glucuronidase and Its Use as a Sweetener in Foods.

Shokuhin Kogyo (Food Ind., Japan), **33**, 46 (1990); Chem. Abstr., **115**, 90834v (1991).

16. JIANG, H.-X., Y. MA, and S.-Q. ZHAO: *Glycyrrhiza* Sweeteners. Shipin Kexue (Food Science, China), **83**, 10 (1986); Chem. Abstr., **106**, 154836w (1987).

17a. KAWAGUCHI, Y.: The Inhibitory Effects of Licorice Extracts on Melanogenesis. 1. *In Vitro* Studies. Jpn. J. Dermatol., **102**, 679 (1992); Chem. Abstr., **117**, 143385z (1992).

17b. KAWAGUCHI, Y., K. GOH, Y. KAWA, M. KASHIMA, and M. MIZOGUCHI: The Inhibitory Effects of Licorice Extracts on Melanogenesis. 2. *In Vivo* Studies. Jpn. J. Dermatol., **102**, 689 (1992); Chem. Abstr., **117**, 143386a (1992).

18a. OOYAMA, Y.: Liquiritin for the Removal of Stains from the Skin. Jpn. patent; Chem. Abstr., **117**, 55948e (1992).

18b. SHIMADA, T., and T. YOKOI: Skin-lightening Preparations for Prevention of Freckles and Blemishes. Jpn. patent; Chem. Abstr., **122**, 64016e (1995).

18c. OGAWA, K.: Cosmetics Containing Glabridin and Ingredients from Animal Source. Jpn. patent; Chem. Abstr., **122**, 64005a (1995).

18d. MAGARA, T., Y. SHIBATA, M. NAGANUMA, M. FUKUDA, and R. KAKO: Skin-lightening Preparations Containing Antiinflammatory Agents and Alkoxy-salicylates. Jpn. patent; Chem. Abstr., **123**, 265807h (1995).

19. HARAMOTO, I.: Licorice Extract has an Inhibitory Effect on Melanogenesis and Improves Melasma Other Pigmented Lesions by Its Topical Use. St. Marianna Med. J. (Japan), **22**, 941 (1994); Chem. Abstr., **123**, 25299a (1995).

20. OKABE, S., and H. OHTSUKI: New Drugs of Peptic Ulcer, Pirenzepine, Aspalon, and Cetraxate. Farumashia (J. Pharmac. Soc. Japan), **15**, 802 (1979).

21. DEPARTMENT OF PHARMACEUTICAL SCIENCES, ROYAL PHARMACEUTICAL SOCIETY OF GREAT BRITAIN. Martindale the Extra Pharmacopoeia, 3rd edn. (REYNOLDS, J.E.F., ed.), pp. 758, 873, 888, London: The Pharmaceutical Press, 1993.

22a. TANAKA, S., A. TAKAHASHI, K. ONODA, K. KAWASHIMA, S. NAKAURA, S. NAGAO, Y. OHNO, T. KAWANISHI, Y. NAKAJI, K. KOBAYASHI, S. SUZUKI, K. NAITO, O. UCHIDA, K, YASUHARA, K. TAKADA, M. SAITO, S. SEKITA, Y. OZAKI, H. SUZUKI, A. TAKANAKA, M. TOBE, and M. HARADA: Toxicological Studies on Biological Effects of the Herbal Drug Extracts in Rats and Mice. 2. Moutan Bark, *Glycyrrhiza* and *Bupleurum* Root. Yakugaku Zasshi, **106**, 671 (1986).

22b. MIRSALIS, J.C., C.M. HAMILTON, J.E. SCHINDLER, C.E. GREEN, and J.E. DABBS: Effects of Soya Bean Flakes and Liquorice Root Extract on Enzyme Induction and Toxicity in $B6C3F_1$ Mice. Food Chem. Toxicol., **31**, 343 (1993); Chem. Abstr., **119**, 94442w (1993).

23a. TANAKA, M., N. MANO, E. AKAZAI, Y. NARUI, F. KATO, and Y. KOYAMA: Inhibition of Mutagenicity by *Glycyrrhiza* Extract and Glycyrrhizin. J. Pharmacobio.-Dyn. (Japan), **10**, 685 (1987).

23b. NGO, H.N., R.W. TEEL, and B.H.S. LAU: Modulation of Mutagenesis, DNA Binding, and Metabolism of Aflatoxin B_1 by Licorice Compounds. Nutrition Res., **12**, 247 (1992).

23c. ZANI, F., M.T. CUZZONI, M. DAGLIA, S. BENVENUTI, G. VAMPA, and P. MAZZA: Inhibition of Mutagenicity in *Salmonella typhimurium* by *Glycyrrhiza glabra* Extract, Glycyrrhizinic Acid, 18α- and 18β-Glycyrrhetinic Acids. Planta Med., **59**, 502 (1993).

23d. NIIKAWA, M., A.-F. WU, T. SATO, H. NAGASE, and H. KITO: Effects of Chinese Medicinal Plant Extracts on Mutagenicity of Trp-P-1. Nat. Med. (Jpn. J. Pharmacognosy), **49**, 329 (1995); Chem. Abstr., **124**, 76471d (1996).

24a. TAKAGI, K., and Y. ISHII: Peptic Ulcer Inhibiting Properties of a New Fraction from Licorice Root (FM 100). 1. Experimental Peptic Ulcer and General Pharmacology. Arzneim. Forsch., **17**, 1544 (1967).
24b. TAKAGI, K.: The Pharmacological Action of Glycyrrhizae Radix. Gendai Toyo Igaku (J. Traditional Sino-Japanese Med., Japan), **2**, 34 (1981).
25a. NOSE, M., M. ITO, K. KAMIMURA, M. SHIMIZU, and Y. OGIHARA: A Comparison of the Antihepatotoxic Activity between Glycyrrhizin and Glycyrrhetinic Acid. Planta Med., **60**, 136 (1994).
25b. WANG, G.-S., and Z.-W. HAN: The Protective Action of *Glycyrrhiza* Flavonoids against Carbon Tetrachloride Hepatotoxicity in Mice. Yaoxue Xuebao (Acta Pharm. Sinica, China), **28**, 572 (1993); Chem. Abstr., **120**, 95683h (1994).
26. WANG, Z.-Y., R. AGARWAL, W.A. KHAN, and H. MUKHTAR: Protection against Benzo[a]pyrene- and N-nitrosodiethylamine-Induced Lung and Forestomach Tumorigenesis in A/J Mice by Water Extracts of Green Tea and Licorice. Carcinogenesis, **13**, 1491 (1992).
27a. TOORS, F.A., and J.I.B. HERCZOG: Acid Production from a Nonsugar Licorice and Different Sugar Substitutes in *Streptococcus mutans* Monoculture and Pooled Plaque-Saliva Mixtures. Caries Res., **12**, 60 (1978); Chem. Abstr., **88**, 101498r (1978).
27b. EDGAR, W.M.: Reduction in Enamel Dissolution by Liquorice and Glycyrrhizinic Acid. J. Dent. Res., **57**, 59 (1978).
27c. NAMBA, T., M. TSUNEZUKA, D.M.R.B. DISSANAYAKE, U. PILAPITIYA, K. SAITO, N. KAKIUCHI, and M. HATTORI: Studies on Dental Caries Prevention by Traditional Medicines. 7. Screening of Ayurvedic Medicines for Anti-Plaque Action. Shoyakugaku Zasshi (Jpn. J. Pharmacognosy), **39**, 146 (1985); Chem. Abstr., **104**, 31601s (1986).
27d. SUIDO, H., T. MAKINO, Y. YAMANE, and H. NAESHIRO: Dentifrice Containing Substances Extracted from Licorice and Analogous Plants. Jpn. patent; Chem. Abstr., **117**, 157447h (1992).
28. ISSHIKI, K., T. NISHINOMIYA, N. NOZAKA, and K. TOKUOKA: Growth Inhibition of Microorganisms by Plant Extracts. Nippon Shokuhin Kogyo Gakkaishi (J. Jpn. Soc. Food Sci. Tech.), **40**, 525 (1993); Biol. Abstr., **96**, 86549 (1993).
29. YONEDA, K.: Production and Resources of Licorice. J. Traditional Sino-Japanese Med. (Japan), **14**, 74 (1993).
30. SATAKE, M.: *Glycyrrhiza* plants. In: Records of Lecture in 2nd Meeting on Crude Drugs; Licorice, p. 1. Tokyo: Tokyo Soc. Pharmacognosy, 1986.
31. INTERNATIONAL LEGUME DATABASE AND INFORMATION SERVICE and CHAPMAN and HALL CHEMICAL DATABASE: Phytochemical Dictionary of the Leguminosae, vol. 1. Plants and their Constituents. (BISBY, F.A., J. BUCKINGHAM, and J.B. HARBORNE, eds.), pp. 338–344. London: Chapman and Hall, 1994.
32. AMIROVA, G.S., and N.P. KIR'YALOV: On Chemotaxonomy of *Glycyrrhiza* L. and *Meristotropis* Fisch. *et* Mey. Gen. Zhurn. Obshchei Biol. (USSR), **50**, 184 (1989); Chem. Abstr., **111**, 36668y (1989).
33. LI, P.-C.: Two New Species of the Genus *Glycyrrhiza* Linn. (Leguminosae) from China. Xi Bei Zhi Wu Yan Jiu (Act. Bot. Bor.-Occ. Sinica, China), **4**, 117 (1984).
34. JIANG SU XIN YI XUE YUAN: Zhong Yao Da Cidaian (Dictionary of Chinese Crude Drugs), vol. 1, pp. 567–573. Shanghai: Shanghai Ke Xue Ji Shu Chu Ban She, 1977.
35. ZENG, L., S.-H. LI, and Z.-C. LOU: Morphological and Histological Studies of Chinese Licorice. Yaoxue Xuebao (Acta Pharm. Sinica, China), **23**, 200 (1988); Biol. Abstr., **86**, 72912 (1988).

36. ZENG, L., Z.-C., LOU, and R.-Y. ZHANG: Quality Evaluation of Chinese Licorice. Yaoxue Xuebao (Acta Pharm. Sinica, China), **26**, 788 (1991); Chem. Abstr., **116**, 46404b (1992).

37. THE DEPARTMENT OF STATISTICS, THE TARIFF BUREAU OF THE FINANCE MINISTRY: Import Data of Foodstuff. Gekkan Fudo Kemikaru (J. Food Chemical, Japan), **15–17**, No. 3 (1994–1996).

38. MIYAZAWA, M., and H. KAMEOKA: Volatile Flavor Components of Glycyrrhizae Radix (*Glycyrrhiza glabra* L. var. *glandulifera* Regel *et* Herder) from China. Flavor Fragrance J., **5**, 157 (1990); Biol. Abstr., **90**, 139187 (1990).

39. SAKAGAMI, H., J. ISEDA, M. KUSAMA, and Y. ISHIZU: Volatile Components of Licorice Roots Produced in Different Countries. Nippon Shokuhin Kogyo Gakkaishi (J. Food Ind. Jpn.), **39**, 257 (1992); Chem. Abstr., **117**, 147229z (1992).

40. SAKAGAMI, H., K. HAYASHI, M. KUSAMA, and Y. ISHIZU: Improvement of Quality of Licorice Extract by Heat Treatment. Nippon Shokuhin Kogyo Gakkaishi (J. Food Ind. Jpn.), **39**, 976 (1992); Chem. Abstr., **118**, 232172w (1993).

41. ZOLOTNITSKAYA, S.YA.: New Alkaloid-Bearing Plants of the Armenian Flora. Izvest. Akad. Nauk. Armyan SSR, Biol. i Sel'khoz. Nauki, **7**, 27 (1954); Chem. Abstr., **48**, 11727d (1954).

42. HAN, Y.-N., M.-S. CHUNG, T.-H. KIM, and B.-H. HAN: Two Tetrahydroquinoline Alkaloids from *Glycyrrhiza uralensis*. Arch. Pharm. Res., **13**, 101 (1990).

43. HAN, Y.-N., and M.-S. CHUNG: A Pyrrolo-pyrimidine Alkaloid from *Glycyrrhiza uralensis*. Arch. Pharm. Res., **13**, 103 (1990).

44. BENETTI, G., and R. RENZETTI: Alkylpyrazines: Important Aroma Components of Licorice Extracts. Ind. Aliment. (Pinerolo, Italy), **33**, 639 (1994); Chem. Abstr., **122**, 29983w (1995).

45. HU, J.-F., and F.-J. SHEN: Structure of a New Alkaloid from the Roots of *Glycyrrhiza yunnanensis*. Gaodeng Xuexiao Huaxue Xuebao (Chem. J. Chin. Univ.), **16**, 1245 (1995); Chem. Abstr., **124**, 50632g (1996).

46. DZHUMAMURATOVA, A., E. SEITMURATOV, D.A. RAKHIMOV, and Z.F. ISMAILOV: Polysaccharides of Some Species of *Glycyrrhiza*. Khim. Prir. Soedin., **1978**, 513; Chem. Nat. Comp. (Engl. transl.), **14**, 437 (1979).

47a. SHI, Y., and G.-Z. YANG: Effect of GPS on Stimulating Murine Lymphocyte Proliferation. Zhongguo Mianyixue Zazhi (China), **2**, 295 (1986); Chem. Abstr., **107**, 75921n (1987).

47b. CHANG, Y.-P., W.-X. BI, and G.-Z. YANG: Studies on the Antiviral Effect of Polysaccharide (GPS) from *Glycyrrhiza uralensis* Fisch. Zhongguo Zhongyao Zazhi (China), **14**, 236 (1989); Chem. Abstr., **111**, 17206e (1989).

48. TOMODA, M., N. SHIMIZU, M. KANARI, R. GONDA, S. ARAI, and Y. OKUDA: Characterization of Two Polysaccharides Having Activity on the Reticuloendothelial System from the Root of *Glycyrrhiza uralensis*. Chem. Pharm. Bull. (Japan), **38**, 1667 (1990).

49. SHIMIZU, N., M. TOMODA, M. KANARI, R. GONDA, A. SATOH, and N. SATOH: A Novel Neutral Polysaccharide Having Activity on the Reticuloendothelial System from the Root of *Glycyrrhiza uralensis*. Chem. Pharm. Bull. (Japan), **38**, 3069 (1990).

50. MESTECHNIKA, N.M., K. DOVLETMURADOV, and V.D. SHCHERBUKHIN: Seed Galactomannan in the Common Licorice (*Glycyrrhiza glabra*). Prikl. Biokhim. Mikrobiol., **27**, 435 (1991); Applied Biochem. Microbiol. (Engl. transl.), **27**, 331 (1991).

51. ZHAO, J.-F., H. KIYOHARA, X.-B. SUN, T. MATSUMOTO, J.-C. CYONG, H. YAMADA, N. TAKEMOTO, and H. KAWAMURA: *In Vitro* Immunostimulating Polysaccharide Fractions from Roots of *Glycyrrhiza uralensis* Fisch. *et* D.C. Phytother. Res., **5**, 206 (1991).

52. J.-F. ZHAO, H. KIYOHARA, H. YAMADA, N. TAKEMOTO, and H. KAWAMURA: Heterogeneity and Characterization of Mitogenic and Anti-Complementary Pectic Polysaccharides from the Roots of *Glycyrrhiza uralensis* Fisch *et* D.C. Carbohyd. Res., **219**, 149 (1991).
53. LIU, B.-C., and J.-N. FANG: Isolation, Purification and Chemical Structure of a Glucan from *Glycyrrhiza uralensis* Fisch. Yaoxue Xuebao (Acta Pharm. Sinica, China), **26**, 672 (1991); Chem. Abstr., **116**, 10278e (1992).
54. SHIMIZU, N., M. TOMODA, M. SATOH, R. GONDA, and N. ÔHARA: Characterization of a Polysaccharide Having Activity on the Reticuloendothelial System from the Stolon of *Glycyrrhiza glabra* var. *glandulifera*. Chem. Pharm. Bull. (Japan), **39**, 2082 (1991).
55. YAMADA, H., and H. KIYOHARA: Antiallergic Polysaccharides Isolated from Plant Roots and Drug Formulation Containing the Same. Jpn. patent; Chem. Abstr., **117**, 29140r (1992).
56. YAMADA. H., H. KIYOHARA, N. TAKEMOTO, J.-F. ZHAO, H. KAWAMURA, Y. KOMATSU, J.-C. CYONG, M. ABURADA, and E. HOSOYA: Mitogenic and Complement Activating Activities of the Herbal Components of Juzen-Taiho-To. Planta Med., **58**, 166 (1992).
57. SHIMIZU, N., M. TOMODA, K. TAKADA, and R. GONDA: The Core Structure and Immunological Activities of Glycyrrhizan UA, the Main Polysaccharide from the Root of *Glycyrrhiza uralensis*. Chem. Pharm. Bull. (Japan), **40**, 2125 (1992).
58. TAKADA, K., M. TOMODA, and N. SHIMIZU: Core Structure of Glycyrrhizan GA, the Main Polysaccharide from the Stolon of *Glycyrrhiza glabra* var. *glandulifera*; Anti-Complementary and Alkaline Phosphatase-Inducing Activities of the Polysaccharide and Its Degradation Products. Chem. Pharm. Bull. (Japan), **40**, 2487 (1992).
59. TOMODA, T., K. TSUCHIYA, N. SHIMIZU, and N. ÔHARA: Two Polysaccharides with Biological Activities from the Stolon of *Glycyrrhiza glabra* var. *glandulifera*. Pharm. Pharmacol. Letters, **4**, 36 (1994); Chem. Abstr., **122**, 177862w (1995).
60. HIRANO, M., T. MATSUMOTO, H. KIYOHARA, and H. YAMADA: Lipopolysaccharide-Independent Limulus Amebocyte Lysate Activating, Mitogenic and Anti-Complementary Activities of Pectic Polysaccharides from Chinese Herbs. Planta Med., **60**, 248 (1994).
61. HIRANO, M., H. KIYOHARA, and H. YAMADA: Existence of a Rhamnogalacturonan II-like Region in Bioactive Pectins from Medicinal Herbs. Planta Med., **60**, 450 (1994).
62. KIYOHARA, H., N. TAKEMOTO, J.-F. ZHAO, H. KAWAMURA, and H. YAMADA: Pectic Polysaccharides from Roots of *Glycyrrhiza uralensis*: Possible Contribution of Neutral Oligosaccharides in the Galacturonase-Resistant Region to Anti-Complementary and Mitogenic Activities. Planta Med., **62**, 14 (1996).
63. HIGUCH, K., K. KAMIBABA, O. MATSUMOTO, M. CHO, Y. OHTSUKA, and Y. MATSUMOTO: Polyamines of Oriental Drugs. Wakan Iyaku Gakkaishi (J. Traditional Jpn.-Chin. Med. Pharm. Soc., Japan), **2**, 250 (1985); Chem. Abstr., **104**, 39570b (1986).
64. ZAYED, S.M.A., A. HASSAN, and M.I. ELGHAMRY: Estrogenic Substances from Egyptian *Glycyrrhiza glabra*. 2. β-Sitosterol as an Estrogenic Principle. Zentr. Veterinaermed., **A11**, 476 (1964); Chem. Abstr., **61**, 14976a (1964).
65. JAIN, S.C., R. KAMAL, and A.K. RATHORE: A Note on Phytosterols in Some Species. Indian Drugs., **17**, 145 (1980); Chem. Abstr., **93**, 66139d (1980).
66. HAYASHI, H., H. FUKUI, and M. TABATA: Distribution Pattern of Saponins in Different Organs of *Glycyrrhiza glabra*. Planta Med., **59**, 351 (1993).

67. Hu, J.-F., Z.-L. Ye, and F.-J. Shen: New Triterpenoidal Sapogenins from the Roots of *Glycyrrhiza yunnanensis*. Yaoxue Xuebao (Acta Pharm. Sinica, China), **30**, 27 (1995); Chem. Abstr., **123**, 22884p (1995).

68a. Gorunovic, M., D. Stosic, D. Runjajic-Antic, and P. Lukic: *Glycyrrhiza echinata* Seed and Seed Oil. Plant Med. Phytother., **17**, 161 (1983); Chem. Abstr., **101**, 51726c (1984).

68b. Shirinyan, E.A., A.G. Panosyan, M.L. Barikyan, and O.M. Avakyan: 9,12,13-Trihydroxy-10-octadecanoic and 9,12,13-Trihydroxy-10,11-epoxyoctadecanoic Acids. New Anti-stressor Compounds from Licorice. Izv. Akad. Nauk SSSR, Ser. Biol., **1988**, 932; Chem. Abstr., **110**, 108136c (1989).

69a. Nishi, H., and I. Morishita: Studies on the Components of Licorice Root Used for Tobacco Flavoring. 1. Fractionation of the Substances in Licorice Root Effective in Improving the Tobacco Smoking Quality. Nippon Nogei Kagaku Kaishi (J. Agric. Chem. Soc. Jpn.), **45**, 507 (1971); Chem. Abstr., **76**, 151016e (1972).

69b. Rudolph, A., R. Becker, G. Scholz, Z. Prochazka, J. Toman, T. Macek, and V. Herout: The Normalizing Factor for the Tomato Mutant "Chloronerva". 22. The Occurrence of the Amino Acid Nicotiamine in Plants and Microorganisms. A Reinvestigation. Biochem. Physiol. Pflanz., **180**, 557 (1985); Chem. Abstr., **103**, 102083s (1985).

70. Russo, G.: Triterpenes of *Glycyrrhiza glabra* L. Fitoterapia, **38**, 98 (1967); Chem. Abstr., **69**, 93590e (1968).

71. Kumagai, A.: Hormone-Like Action of Licorice and Glycyrrhizin. Taisha (Metabolism Disease, Japan), **10**, 632 (1973).

72. Watanabe, K.: Pharmacology of Licorice Root. Taisha (Metabolism Disease, Japan), **10**, 626 (1973).

73. Oura, H., S. Nakashima, A. Kumagai, and M. Takata: Immunosuppressive Effect of Licorice Root Extracts. Taisha (Metabolism Disease, Japan), **10**, 651 (1973).

74. Shibata, S., and T. Saitoh: The Chemical Constituents of Licorice Roots. Taisha (Metabolism Disease, Japan), **10**, 619 (1973); Chem. Abstr., **81**, 101792n (1974).

75. Yamamoto, M.: Liquorice and Lipid Metabolism. Taisha (Metabolism Disease, Japan), **10**, 646 (1973).

76. Yano, S.: Clinical Application of Licorice and Its Historical Background. Taisha (Metabolism Disease, Japan), **10**, 659 (1973).

77. Baltassat-Millet, F., and S. Ferry: Pharmacology and Toxicology of Licorice, *Glycyrrhiza glabra* L. Lyon Pharm., **27**, 49 (1976); Chem. Abstr., **84**, 144362b (1976).

78. Shibata, S., and T. Saitoh: Flavonoid Compounds in Licorice Root. J. Indian Chem. Soc., **55**, 1184 (1978).

79. Liu, G.-J.: Xinjiang Gan Cao (Xinjiang Licorice). Urumchi: Xinjiang People's Publishing House (China), 1982.

80. Nisimoto, K., and I. Yasuda: The Quality of *Glycyrrhiza*. Gendai Toyo Igaku (J. Traditional Sino-Japanese Med., Japan), **2**, 56 (1981); Chem. Abstr., **94**, 214409e (1981).

81. Shibata, S.: Chemistry of Licorice Root. Gendai Toyo Igaku (J. Traditional Sino-Japanese Med., Japan), **2**, 46 (1981); Chem. Abstr., **94**, 214408d (1981).

82. Lutomski, J.: Chemistry and Therapeutic Use of Licorice (*Glycyrrhiza glabra* L.). Pharm. unserer Zeit, **12**, 49 (1983).

83. Shibata, S.: A Retrospect on the Studies of Naturally Occurring Medicinal Materials. Shoyakugaku Zasshi (Jpn. J. Pharmacognosy), **40**, 1 (1986); Chem. Abstr., **105**, 84992x (1986).

84. CAI, L., R.-Y. ZHANG, and J.-H. ZHANG: Advance in Triterpenoids from *Glycyrrhiza* Plants. Tianran Chanwu Yanjiu Yu Kaifa (Nat. Prod. Research Develop., China), **1** (1), 47 (1989); Chem. Abstr., **115**, 275684q (1991).

85. LIU, Q., and Y.-L. LIU: Flavonoids from *Glycyrrhiza* genus. Zhongguo Yaoxue Zazhi (Chin. Pharm. Bull.), **24**, 705 (1989); Chem. Abstr., **113**, 20844h (1990).

86. AYABE, S.: Regulation of Flavonoid Biosynthesis in Cultured *Glycyrrhiza echinata* Cells. Shokubutu Soshiki Baiyo (Plant Tissue Culture Letters, Japan), **6**, 113 (1989); Chem. Abstr., **113**, 130602k (1990).

87. BIELENBERG, J.: Licorice can also Induce Intoxication. Pharm. Ztg., **49**, 9 (1989); Chem. Abstr., **112**, 71703y (1990).

88. KIKUTI, H.: Licorice Extract as a Sweetening Agent with Useful Functions. Gekkan Fudo Kemikaru (J. Food Chemistry, Japan), **10**, 79 (1989); Chem. Abstr., **112**, 34463z (1990).

89. KITAGAWA, I.: Chemical Investigation of Naturally Occurring Drug Materials. Elucidation of Scientific Basis for Traditional Medicines and Exploitation of New Naturally Occurring Drugs. Yakugaku Zasshi, **112**, 1 (1992).

90. HIAI, S., and T. NAGASAWA: Biochemical and Pharmacological Studies on Licorice. Gendai Toyo Igaku (J. Traditional Sino-Japanese Med., Japan), **14**, 90 (1993).

91. KITAGAWA, I., and K. HORI: Chemical Constituents of Licorice-Roots. Gendai Toyo Igaku (J. Traditional Sino-Japanese Med., Japan), **14**, 80 (1993).

92. BAKER, M.E.: Licorice and Enzymes Other than 11β-Hydroxysteroid Dehydrogenase: An Evolutionary Perspective. Steroids, **59**, 136 (1994).

93. YIN, J., and L.-G. GUO (eds.): Zhong Yao Xian Dai Yan Jiu Yu Lin Chuang Ying Yong (Modern Chinese Crude Drugs and Its Clinical Applications), pp. 196–212. Beijing (China): Xue Yuan, 1993.

94. HU, J.-F., and F.-J. SHEN: The Deep Exploitation and Comprehensive Utilization of Licorice (*Glycyrrhiza* L.) in Medicine and Other Parties Concerned. Zhongcaoyao (Chin. Herbal Med.), **26**, 39 (1995).

95. JIA, S.-S., and G.-F. QIU: Active Constituents and Their Exploitation of the Aerial Part of Licorice. Zhongcaoyao (Chin. Herbal Med.), **25**, 106 (1994); Chem. Abstr., **121**, 5030m (1994).

96. SAITO, K.: Molecular Genetics and Biotechnology in Medicinal Plants: Studies by Transgenic Plants. Yakugaku Zasshi, **114**, 1 (1994).

97. FENG, X.-Z.: Recent Studies on Flavonoids from Chinese Traditional and Herb Drugs. UNESCO Regional Seminar on the Chemistry, Pharmacology and Clinical Use of Flavonoid Compounds, Abstract Papers, p. 72. Taejon (Korea), October 1995.

98. ZENG, L., R.-Y. ZHANG, D. WANG, C.-Y. GAO, and Z.-C. LOU: The Chemical Constituents of *Glycyrrhiza aspera* Root. Zhiwu Xuebao (Acta Bot. Sinica, China), **33**, 124 (1991); Chem. Abstr., **115**, 228388m (1991).

99. KIR'YALOV, N.P.: New Triterpene Acids from *Glycyrrhiza*. Vop. Izuch. Ispol'z. Solodki v SSSR, **1966**, 123; Chem. Abstr., **68**, 66396u (1968).

100. KITAGAWA, I., K. HORI, T. TANIYAMA, J.-L. ZHOU, and M. YOSHIKAWA: Saponin and Sapogenol. 47. On the Constituents of the Roots of *Glycyrrhiza uralensis* Fischer from Northeastern China. 1. Licorice-Saponins A3, B2, and C2. Chem. Pharm. Bull. (Japan), **41**, 43 (1993).

101. RUZICKA, L., and A.G. VAN VEEN: Saponins, Sapogenins and Analogous Compounds. 2. Relationships between Sapogenins, Higher Terpene Compounds and Sterols. Z. Physiol. Chem., **184**, 69 (1929); Chem. Abstr., **24**, 856 (1930).

102. BEOVIDE, D.S.: Licorice Extract and Glycyrrhizic and Glycyrrhetic Acid. Rev. Acad. Cienc. Madrid., **30**, 555 (1933); Chem. Abstr., **28**, 2719[6] (1934).

103. RUZICKA, L., and H. LEUENBERGER: Polyterpenes and Polyterpenoids. 109. Glycyrrhetic Acid. Helv. Chim. Acta, **19**, 1402 (1936).

104. VOSS, W., and J. PFIRSCHKE: Glycyrrhizin. 2. A Novel Disaccharide as the Sugar Component of Glycyrrhizin. Chem. Ber., **70**, 132 (1937).

105. RUZICKA, L., M. FURTER, and H. LEUENBERGER: Polyterpenes and Polyterpenoids. 111. Empirical Formula of Glycyrrhetic Acid. Helv. Chim. Acta, **20**, 312 (1937).

106. BERGMANN, E., and F. BERGMANN: Glycyrrhetic Acid. Helv. Chim. Acta, **20**, 207 (1937).

107. VOSS, W., P. KLEIN, and H. SAUER: Glycyrrhizin. Chem. Ber., **70**, 122 (1937).

108. VOSS, W., and G. BUTTER: Glycyrrhizin. 3. Isomerism of the Glycyrrhetinic Acids. Chem. Ber., **70**, 1212 (1937).

109. RUZICKA, L., and S.L. COHEN: Polyterpenes and Polyterpenoids. 113. Oxidation in the Oleanolic Acid Series without Rupture of the Ring System. Nature of the Fourth Oxygen Atom of Glycyrrhetic Acid. Helv. Chim. Acta, **20**, 804 (1937).

110. RUZICKA, L., H. LEUENBERGER, and H. SCHELLENBERG: Polyterpenes and Polyterpenoids. 118. Catalytic Hydrogenation of the α,β-Unsaturated Keto Group in Glycyrrhetic Acid and Keto-α-amyrin. Helv. Chim. Acta, **20**, 1271 (1937).

111. KARIYONE, T., and O. NONAKA: On the Sweet Constituents of Liquorice Root. 1. Yakugaku Zasshi, **57**, 166 (1937).

112. TAKEDA, K.: Glycyrrhetic Acid. Yakugaku Zasshi, **58**, 731 (1938).

113. KURONO, G.: On the Sweet Constituents of Liquorice Root. Yakugaku Zasshi, **58**, 770 (1938).

114. RUZICKA, L., and A. MARXER: Studies of Triterpene. 44. Conversion of Glycyrrhetinic Acid into β-Amyrin. Helv. Chim. Acta, **22**, 195 (1939).

115. VIRATELLE, R.: The Constitution of Glycyrrhizic Acid. Bull. Sci. Pharmacol., **45**, 346 (1938); Chem. Abstr., **33**, 2528[1] (1939).

116. GIACOMELLO, G.: Structure of Various Triterpenes. Atti. Acad. Lincei, Classe Sci. Fis. Mat. Nat., **27**, 574 (1938); Chem. Abstr., **33**, 1567[8] (1939).

117. BILHAM, P., G.A.R. KON, and W.C.J. ROSS: Sapogenins. 14. The Constitution of Glycyrrhetic Acid and Its Relation to Oleanolic Acid. J. Chem. Soc. (London), **1942**, 535.

118. KON, G.A.R., and W.C.J. ROSS: Sapogenins. 17. The Position of the Carboxyl Group in Oleanolic and Glycyrrhetic Acids. J. Chem. Soc. (London), **1942**, 741.

119. RUZICKA, L., and O. JEGER: Studies of Triterpenes. 67. Position of the Carbonyl Group in Glycyrrhetinic Acid. Helv. Chim. Acta, **25**, 775 (1942).

120. RUZICKA, L., O. JEGER, and M. WINTER: Studies of Triterpene. 75. The Position of the Carbonyl Group in Oleanolic Acid and Glycyrrhetinic Acid. Helv. Chim. Acta, **26**, 265 (1943).

121. RUZICKA, L., O. JEGER, and W. INGOLD: Studies of Triterpenes. 84. New Proof for the Difference in Position of the Carboxylic Acid Group in Oleanolic and Glycyrrhetinic Acids. Helv. Chim. Acta, **26**, 2278 (1943).

122. JEGER, O., J. NORYMBERSKI, and L. RUZICKA: Studies of Triterpenes. 90. An Oxidation Product $C_{32}H_{46}O_5$ from Acetyl-β-amyrin and Two Isomeric Oxidation Products $C_{33}H_{46}O_7$ from Methyl Acetyloleanolate and Methyl Acetylglycyrrhetinate. Helv. Chim. Acta, **27**, 1532 (1944).

123. LYTHGOE, B., and S. TRIPPETT: The Constitution of the Disaccharide of Glycyrrhinic Acid. J. Chem. Soc. (London), **1950**, 1983.

124. BEATON, J.M., and F.S. SPRING: Triterpenoids. 42. The Configuration of the Carboxyl Group in Glycyrrhetic Acid. J. Chem. Soc. (London), **1955**, 3126.
125. MURAV'EV, I.A., V.D. PONOMAREV, and É.K. KRAVCHENKO: The Structure of the Monoammonium Salt of Glycyrrhizic Acid. Khim. Prir. Soedin., **1971**, 121; Chem. Nat. Comp. (Engl. transl.), **7**, 114 (1973).
126a. KHALILOV, L.M., L.A. BALTINA, L.V. SPIRIKHIN, E.V. VASIL'EVA, R.M. KONDRATENKO, A.A. PANASENKO, and G.A. TOLSTIKOV: ^{13}C NMR Spectra of Biologically Active Compounds. 8. Stereochemistry of a Triterpene Glycoside, Glycyrrhizic Acids, and Its Derivatives. Khim. Prir. Soedin., **1989**, 500; Chem. Nat. Comp. (Engl. transl.), **25**, 426 (1990).
126b. SHIBATA, S.: Antitumor-promoting and Anti-inflammatory Activities of Licorice Principles and Their Modified Compounds. In: ACS Symposium Series 547, Food Phytochemicals for Cancer Prevention II. Teas, Spices, and Herbs (HO, C.-T., T. OSAWA, M.-T. HUANG, and R. T. ROSEN, eds.), pp 308–321. Washington D.C.: American Chemical Society, 1994.
127. ZHANG, R.-Y., J.-H. ZHANG, and M.-T. WANG: Studies on the Saponins from the Roots of *Glycyrrhiza uralensis* Fisch. Yaoxue Xuebao (Acta Pharm. Sinica, China), **21**, 510 (1986); Chem. Abstr., **105**, 187603a (1986).
128. LIU, Q., and Y.-L. LIU: Chemical Constituents of *Glycyrrhiza eurycarpa* P. C. Li. 2. Zhiwu Xuebao (Acta Bot. Sinica, China), **33**, 314 (1991); Chem. Abstr., **116**, 102644u (1992).
129. AYABE, S., H. TAKANO, T. FUJITA, T. FURUYA, H. HIROTA, and T. TAKAHASHI: Studies on Plant Tissue Cultures. 72. Triterpenoid Biosynthesis in Tissue Cultures of *Glycyrrhiza glabra* var. *glandulifera*. Plant Cell Rep., **9**, 181 (1990).
130. HAYASHI, H., H. FUKUI, and M. TABATA: Examination of Triterpenoids Produced by Callus and Cell Suspension Cultures of *Glycyrrhiza glabra*. Plant Cell Rep., **7**, 508 (1988).
131. ARIAS-CASTRO, C., A.H. SCRAGG, and M.A. RODRIGUEZ-MENDIOLA: The Effect of Cultural Conditions on the Accumulation of Formononetin by Suspension Cultures of *Glycyrrhiza glabra*. Plant Cell, Tissue Organ Cult., **34**, 63 (1993).
132. HENRY, M., A.-M. EDY, and B. MARTY: Isolation of Licorice Protoplasts (*Glycyrrhiza glabra* L. var. *typica* Reg. et Hed.) from Cell Suspension Cultures Not Producing Glycyrrhetinic Acid. C.R. Acad. Sci., Ser. 3, **299**, 899 (1984); Chem. Abstr., **102**, 182523c (1985).
133. WU, C.-H., E.M. ZABAWA, and P.M. TOWNSLEY: The Single Cell Suspension Culture of the Licorice Plant, *Glycyrrhiza glabra*. Can. Inst. Food. Sci. Tech. J., **7**, 105 (1974).
134. SAKAMOTO, K., K. IIDA, and T. KOYANO: Glycyrrhizin Manufacture by Tissue Culture of Licorice. Jpn. patent; Chem. Abstr., **114**, 80082p (1990).
135. HIMI, M., K. FUJITA, and S. YONETANI: Glycyrrhizin Manufacture Enhancement by Licorice with Physical Stress. Jpn. patent; Chem. Abstr., **117**, 149467z (1992).
136. TAMAKI, E., I. MORISHITA, K. NISHIDA, K. KATO, and T. MATSUMOTO: Licorice-like Products as Tobacco Aromatics. Ger. Patent; Chem. Abstr., **77**, 85879f (1972).
137. FUJITA, Y., K. TERANISHI, and T. FURUKAWA: Glycyrrhizin. Jpn. Patent; Chem. Abstr., **89**, 194145g (1978).
138. HAYASHI, H., T. SAKAI, H. FUKUI, and M. TABATA: Formation of Soyasaponins in Licorice Cell Suspension Cultures. Phytochem., **29**, 3127 (1990).
139. KO, K.-S., H. NOGUCHI, Y. EBIZUKA, and U. SANKAWA: Oligoside Production by Hairy Root Cultures Transformed by Ri Plasmids. Chem. Pharm. Bull. (Japan), **37**, 245 (1989).

140. Kojima, K.: Natural Sweetening. Shoku no Kagaku (Food Sci., Japan), **56**, 40 (1980).

141. Onishi, T.: Studies again to Sweets Used in Soy Sauce. Chomi Kagaku (Seasoning Sci., Japan), **17**, 31 (1970); Chem. Abstr., **76**, 2644f (1972).

142. Kuramoto, T., Y. Ito, M. Oda, Y. Tamura, and S. Kitahara: Microbial Production of Glycyrrhetic Acid 3-*O*-Mono-β-D-glucuronide from Glycyrrhizin by *Cryptococcus magnus* MG-27. Biosci. Biotech. Biochem. (Japan), **58**, 455 (1994).

143. Mizutani, K., T. Kuramoto, Y. Tamura, N. Ohtake, S. Doi, M. Nakaura, and O. Tanaka: Sweetness of Glycyrrhetic Acid 3-*O*-β-D-Monoglucuronide and the Related Glycosides. Biosci. Biotech. Biochem. (Japan), **58**, 554 (1994).

144. Swiss Pharmaceutical Society: International Drug Directory 1992/1993, pp. 201, 430. Stuttgart: Medpharm Scientific Publishers, 1992.

145. Japan Pharmaceutical Information Center: Drugs in Japan, Ethical Drugs, 1995–August, pp. 21, 384, 385, 944. Tokyo: Yakugyo Jiho, 1995.

146. Suzuki, H., Y. Ohta, T. Takino, K. Fujisawa, C. Hirayama, N. Shimizu, and Y. Aso: Effects of Stronger Neo-Minophagen C in Chronic Hepatitis, Assessment with Double Blind Test. Igaku no Ayumi (J. Clinical Exper. Med., Japan), **102**, 562 (1977).

147. Sekiguchi, S., S. Sakamoto, M. Kaneko, E. Sasaki, and Y. Kasai: Studies of Precaution against Transfusion Hepatitis with Intravenous of Stronger Neo-Minophagen C at High Doses. Gendai Iryou (Modern Therapy, Japan), **14**, 341 (1982).

148. Ito, M., H. Nakashima, M. Baba, R. Pauwels, E.D. Clercq, S. Shigeta, and N. Yamamoto: Inhibitory Effect of Glycyrrhizin on the *in vitro* Infectivity and Cytopathic Activity of the Human Immunodeficiency Virus [HIV (HTLV-III/LAV)]. Antiviral Res., **7**, 127 (1987).

149. Gotoh, Y., K. Tada, K. Yamada, M. Minamitani, M. Negishi, M. Fujimaki, S. Ikematsu, M. Hada, K. Mori, M. Ito, S. Shigeta, H. Nakashima, N. Yamamoto, and Y. Shiokawa: Administration of Glycyrrhizin to Patients with Human Immunodeficiency Virus Infection. Igaku no Ayumi (J. Clinical Exper. Med., Japan), **140**, 619 (1987).

150. Mori, K., H. Sakai, S. Suzuki, K. Tutui, Y. Akutsu, M. Ishikawa, Y. Seino, Y. Satoh, Y. Gotou, T. Uchida, S. Kariyone, Y. Endo, and A. Miura: Effects of Intravenous Glycyrrhizin (SNMC: Stronger Neo-Minophagen C®) at High Dose for HIV Carrier in Hemophilia Patients. 1. Rinsho to Kenkyu (Jpn. J. Clin. Exp. Med.), **65**, 981 (1988).

151. Mori, K., H. Sakai, S. Suzuki, K. Sugai, Y. Akutsu, M. Ishikawa, Y. Seino, N. Ishida, T. Uchida, S. Kariyone, Y. Endo, and A. Miura: Effects of Glycyrrhizin (SNMC: Stronger Neo-Minophagen C®) in Hemophilia Patients with HIV Infection. Tohoku J. Exp. Med. (Japan), **158**, 25 (1989); Biol. Abstr., **88**, 82439 (1989).

152. Mori, K., H. Sakai, S. Suzuki, Y. Akutsu, M. Ishikawa, M. Aihara, M. Yokoyama, Y. Sato, S. Okaniwa, Y. Endo, H. Sasaki, S. Saito, T. Hayashi, K. Niikawa, S. Suzuki, T. Uchida, and K. Hiwatashi: The Present Status in Prophylaxis and Treatment of HIV Infected Patients with Hemophilia in Japan. Rinsho Byori (Jpn. J. Clin. Pathol.), **37**, 1200 (1989).

153. Endo, Y., S. Mamiya, K. Iwamoto, H. Niitsu, T. Ito, and A. Miura: The Frequency of the Patients with Positive HIV-Antibody in the Various Groups and the Prognosis of these Patients Treated with Glycyrrhizin (SNMC). Rinsho to Yakuri (Jpn. J. Clin. Pathol.), **38**, 188 (1990).

154. MORI, K., H. SAKAI, S. SUZUKI, K. SUGAI, Y. AKUTSU, M. ISHIKAWA, Y. SEINO: Effects of High-Dose Glycyrrhizin (SNMC: Stronger Neo-Minophagen C®) in Hemophilia Patients with HIV Infection. 2. Rinsho to Kenkyu (Jpn. J. Clin. Exp. Med.), **66**, 1653 (1989).

155. JAMES, J.S.: Berlin Conference: HIV Miscellaneous, Part I. AIDS Treatment News, No. 181 (1993).

156. MORI, K., H. SAKAI, S. SUZUKI, Y. AKUTSU, M. ISHIKAWA, K. TADA, M. AIHARA, Y. SAWADA, M. YOKOYAMA, Y. SATO, Y. ENDO, Z. SUZUKI, S. SATO, H. SASAKI, S. YOKOYAMA, T. HAYASHI, T. UCHIDA, K. HIWATASHI, N. ISHIDA, M. FUJIMAKI, and K. YAMADA: Effects of Glycyrrhizin (SNMC: Stronger Neo-Minophagen C®) in Hemophilia Patients with HIV-1 Infection. J. Naturopathic Med., **4**, 2 (1993).

157. SMITH, D.: *Glycyrrhiza*: Research Still Promising, Still Limited. Aids Treatment News No. 103, 2nd item (1990).

158. ELGAMAL, M.H.A., M.B.E. FAYEZ, and G. SNATZKE: Constituents of Local Planta. 6. Liquoric Acid, a New Triterpenoid from the Roots of *Glycyrrhiza glabra* L. Tetrahedron, **21**, 2109 (1965).

159. CANONICA, L., RUSSO, G, and A. BONATI: Triterpenes of *Glycyrrhiza glabra*. 1. Two New Lactones with an Oleanane Structure. Gazz. Chim. Ital., **96**, 772 (1966).

160. CANONICA, L., G. RUSSO, and E. BOMBARDELLI: Triterpenes of *Glycyrrhiza glabra*. 2. Liquiritic Acid. Gazz. Chim. Ital., **96**, 833 (1966).

161. CANONICA, L., B. DANIELI, P. MANITTO, and G. RUSSO: Triterpenes of *Glycyrrhiza glabra*. 3. Structure of Isoglabrolide. Gazz. Chim. Ital., **96**, 843 (1966).

162. CANONICA, L., B. DANIELI, G. RUSSO, and A. BONATI: Triterpenes from *Glycyrrhiza glabra*. 4. 18α-Hydroxyglycyrrhetic Acid. Gazz. Chim. Ital., **97**, 769 (1967).

163. CANONICA, L., B. DANIELI, P. MANITTO, G. RUSSO, and E. BOMBARDELLI: *Glycyrrhiza glabra* Triterpenes. 5. Glycyrretol and 21α-Hydroxyisoglabrolide. Gazz. Chim. Ital., **97**, 1347 (1967).

164. CANONICA, L., B. DANIELI, P. MANITTO, G. RUSSO, and A. BONATI: *Glycyrrhiza glabra* Triterpenes. 6. 24-Hydroxyglycyrrhetic Acid and 24-Hydroxy-11-deoxyglycyrrhetic Acid. Gazz. Chim. Ital., **97**, 1359 (1967).

165. RUSSO, G.: Triterpenoids. Corsi Semin. Chim., **1968**, 20; Chem. Abstr., **72**, 21799u (1970).

166. CANONICA, L., B. DANIELI, M. BRUNO, R. PAOLO, G. RUSSO, E. BOMBARDELLI, and A. BONATI: Triterpenes of *Glycyrrhiza glabra*. 7. 24-Hydroxyliquiritic Acid [3β,24-Dihydroxy-11-oxoolean-12-en-29-oic acid] and Liquiridiolic Acid [3β,21α,24-Trihydroxyolean-12-en-29-oic acid]. Gazz. Chim. Ital., **98**, 712 (1968).

167. BEATON, J.M., and F.S. SPRING: Triterpenoids. 51. The Isolation and Characterization of Glabric Acid, a New Triterpenoid Acid from Liquorice Root. J. Chem. Soc. (London), **1956**, 2417.

168. ELGAMAL, M.H.A., and M.B.E. FAYEZ: The Structure of Glabric Acid, a Further Triterpenoid Constituent of *Glycyrrhiza glabra* L. Acta Chim. Acad. Sci. Hungar., **58**, 75 (1968); Chem. Abstr., **70**, 29115x (1969).

169. BOGATKINA, V.F., I.A. MURAV'EV, E.F. STEPANOVA, and N.P. KIR'YALOV: Triterpene Compounds from the Epigeal Mass of *Glycyrrhiza glabra*. Khim. Prir. Soedin., **1974**, 101; Chem. Nat. Comp. (Engl. transl.), **10**, 114 (1976).

170. KIR'YALOV, N.P., and T.N. NAUGOL'NAYA: A New Triterpene Hydroxy Oxo Acid, Uralenic Acid, from *Glycyrrhiza uralensis*. Zhurn. Obshchei Khimii (USSR) **34**, 2814 (1964); Chem. Abstr., **61**, 14725e (1964).

171. KIR'YALOV, N.P., and MURAV'EV, I.A.: Triterpenoid Compounds of *Glycyrrhiza glabra* Grass. Khim. Prir. Soedin., **1970**, 770; Chem. Abstr., **74**, 95466w (1971).

172. Belous, V.N., L.G. Matyukhina, and A.A. Ryabinin: Uralenic acid. Zhurn. Obshchei Khimmii (USSR), **35**, 401 (1965); Chem. Abstr., **62**, 13186a (1965).

173. Elgamal, M.H.A., and B.A.H. El-Tawil: Constituents of Local Plants. 18. 28-Hydroxyglycyrrhetic Acid, a New Triterpenoid Isolated from the Roots of *Glycyrrhiza glabra*. Planta Med., **27**, 159 (1975).

174. Shu, Y.-H., Y.-I. Zhao, and R.-Y. Zhang: Isolation and Structural Identification of Triterpene Sapogenins from *Glycyrrhiza uralensis* Fisch. Yaoxue Xuebao (Acta Pharm. Sinica, China), **20**, 193 (1985); Chem. Abstr., **103**, 119932r (1985).

175. Shu, Y.-H., R.-Y. Zhang, Y.-Y. Zhao, J.-W. Zhang, and W.-D. Tong: Isolation and Structure Determination of New Triterpene Sapogenins from *Glycyrrhiza uralensis* Fisch. Yaoxue Xuebao (Acta Pharm. Sinica, China), **22**, 512 (1987); Chem. Abstr., **107**, 214819a (1987).

176. Jia, Q., B. Wang, Y.-H. Shu, R.-Y. Zhang, C.-Y. Gao, L. Qiao, and J.-H. Pang: The Structure of Glyuranolide, a New Triterpene of *Glycyrrhiza uralensis* Fisch. Yaoxue Xuebao (Acta Pharm. Sinica, China), **24**, 348 (1989); Chem. Abstr., **111**, 228990u (1989).

177. Gao, C.-Y., L. Qiao, Q. Jia, and Z.-L. Zhang: Study and Identification of the Chemical Structure of a New Component in *Glycyrrhiza uralensis* Fisch. by NMR Method. Bopuxue Zazhi (Chin. J. Mag. Res.), **7**, 11 (1990); Chem. Abstr., **113**, 55907h (1990).

178. Shen, F.-J., J.-F. Hu, Y.-C. Yu, Z.-D. Xu: Studies on Chemical Constituents of *Glycyrrhiza uralensis* Fisch. Gaodeng Xuexiao Huaxue Xuebao (Chem. J. Chin. Univ.), **16**, 572 (1995); Chem. Abstr., **123**, 138758t (1995).

179. Kir'yalov, N.P., V.F. Bogatkina, and E.Yu. Barkaeva: Triterpenoids of the Roots of *Glycyrrhiza uralensis*. Khim. Prir. Soedin., **1974**, 102; Chem. Nat. Comp. (Engl. transl.), **10**, 112 (1975).

180. Semenchenko, V.F.: Chemical Characteristics of *Glycyrrhiza echinata* Roots. Rast. Resur., **5**, 394 (1969); Chem. Abstr., **72**, 19117b (1970).

181. Kir'yalov, N.P., and T.N. Naugol'naya: A New Triterpene Acid (Macedonic Acid) from *Glycyrrhiza macedonica*. Zhurn. Obshchei Khimii (USSR), **33**, 697 (1963); Chem. Abstr., **59**, 679f (1963).

182. Kir'yalov, N.P., and T.N. Naugol'naya: Triterpene Acid (Echinatic Acid) from the Roots of *Glycyrrhiza echinata*. Zhurn. Obshchei Khimii (USSR), **33**, 700 (1963); Chem. Abstr., **59**, 2867d (1963).

183. Kir'yalov, N.P.: Structure of Macedonic Acid. Khim. Prir. Soedin., **1969**, 448; Chem. Abstr., **72**, 67136d (1970).

184. Kir'yalov, N.P., and V.F. Bogatkina: Isomacedonic Acid from the Roots of *Glycyrrhiza echinata*. Khim. Prir. Soedin., **7**, 123 (1971); Chem. Nat. Comp. (Engl. transl.), **7**, 117 (1971).

185. Zorina, A.D., L.G. Matyukhina, I.A. Saltykova, and A.G. Shavva: The Structures of Macedonic and Isomacedonic Acids. Zhurn. Org. Khim., **9**, 1673 (1973); J. Organ. Chem. (Engl. transl.), **9**, 1696 (1974).

186. Zorina, A.D., L.G. Matyukhina, A.G. Chavva, and L.A. Saltikova: Structure of Macedonic Acid, a Pentacyclic Triterpene. Tetrahedron Letters, **1972**, 1841.

187. Semenchenko, V.F.: Echinatic Acid from the Roots of *Glycyrrhiza macedonica*. Khim. Prir. Soedin., **6**, 490 (1970); Chem. Nat. Comp. (Engl. transl.), **6**, 512 (1970).

188. Kir'yalov, N.P., and V.F. Bogatkina: Structure of Echinatic Acid. Khim. Prir. Soedin., **1969**, 447; Chem. Abstr., **72**, 67137e (1970).

189. Kir'yalov, N.P., and V.F. Bogatkina: Isoechinatic Acid from *Glycyrrhiza echinata* Roots. Khim. Prir. Soedin., **1971**, 378; Chem. Abstr., **75**, 115893j (1971).

190. KIR'YALOV, N.P., and V.F. BOGATKINA: Structure of Echinatic and Isoechinatic Acids. Khim. Prir. Soedin., **1977**, 120; Chem. Abstr., **88**, 7105p (1978).

191. MIRHOM, Y.W., A.G. HANNA, M.H.A. ELGAMAL, K. SZENDREI, and J. REISCH: A Novel Triterpenoid Isolated from the Roots of *Glycyrrhiza echinata* L. Z. Naturforsch., **45b**, 1111 (1990).

192. ZENG, L., R.-Y. ZHANG, P. WEI, D. WANG, C.-Y. GAO, and Z.-C. LOU: New Triterpenoidal Sapogenins from the Roots of *Glycyrrhiza yunnanensis*. Yaoxue Xuebao (Acta Pharm. Sinica, China), **25**, 515 (1990); Chem. Abstr., **114**, 20975x (1991).

193. ZENG, L., R.-Y. ZHANG, D. WANG, and Z.-C. LOU: Two Triterpenoids from Roots of *Glycyrrhiza yunnanensis*. Phytochem., **29**, 3605 (1990).

194. ZENG, L., R.-Y. ZHANG, D. WANG, J.-H. PANG, Z.-L. ZHANG, C.-Y. GAO, and Z.-C. LOU: Glyyunnanprosapogenin and Glyyunnansapogenin from the Roots of *Glycyrrhiza yunnanensis*. Yaoxue Xuebao (Acta Pharm. Sinica, China), **25**, 750 (1990); Chem. Abstr., **117**, 86657p (1992).

195. ZENG, L., R.-Y. ZHANG, D. WANG, Z.-L. ZHANG, and Z.-C. LOU: Glyyunnansapogenins G and H: Two New Pentacyclic Triterpenoids of the 18αH-Oleana-9(11),12-diene Type from *Glycyrrhiza yunnanensis* Roots. Planta Med., **57**, 165 (1991).

196. LIU, J.-H., S.-S. YANG, Y.-Q. FU, C.-L. YUAN, and B. LIU: Studies on Chemical Constituents from *Glycyrrhiza pallidiflora* Maxim. Yaoxue Xuebao (Acta Pharm. Sinica, China), **25**, 689 (1990); Chem. Abstr., **114**, 160672w (1991).

197. MURAV'EV, I.A., V.F. SEMENCHENKO, and L.V. KUKHAREVA: Macedonic Acid from the Roots of *Glycyrrhiza pallidiflora*. Khim. Prir. Soedin., **1971**, 122; Chem. Nat. Comp. (Engl. transl.), **7**, 116 (1973).

198. KAN, Y.-M., H.-B. ZHAO, X.-H. LIU, Y. ZHU, R. XIA, Y.-P. ZHU, and M. LIN: Chemical Constituents of *Glycyrrhiza pallidiflora* (Prickly Fruit Licorice). Zhongcaoyao (Chin. Herbal Med.), **25**, 3 (1994); Chem. Abstr., **121**, 91431x (1994).

199. KAN, Y.-M., and R.-G. WANG: Constituents of *Glycyrrhiza pallidiflora*. Fitoterapia, **65**, 91 (1994); Chem. Abstr., **121**, 153319c (1994).

200. KIR'YALOV, N.P., V.F. BOGATKINA, and T.P. NADEZHINA: Glabrolide from *Glycyrrhiza aspera* Roots. Khim. Prir. Soedin., **1973**, 277; Chem. Abstr., **79**, 40014v (1973).

201. KIR'YALOV, N.P., V.F. BOGATKINA, and T.P. NADEZHINA: 24-Hydroxyglycyrrhetic Acid from the Roots of *Glycyrrhiza korshinskyi*. Khim. Prir. Soedin., **1972**, 395; Chem. Nat. Comp. (Engl. transl.), **8**, 393 (1974).

202. LIANG. H., and R.-Y. ZHANG: Studies on the Triterpenoids from Roots of *Glycyrrhiza squamulosa* Froanch. Yaoxue Xuebao (Acta Pharm. Sinica, China), **28**, 116 (1993); Chem. Abstr., **119**, 91203h (1993).

203. KITAGAWA, I., J.-L. ZHOU, M. SAKAGAMI, T. TANIYAMA, and M. YOSHIKAWA: Licorice-Saponins A3, B2, C2, D3, and E2, Five New Oleanene-Type Triterpene Oligoglycosides from Chinese Glycyrrhizae Radix. Chem. Pharm. Bull. (Japan), **36**, 3710 (1988).

204. KITAGAWA, I., J.-L. ZHOU, M. SAKAGAMI, E. UCHIDA, and M. YOSHIKAWA: Licorice-saponins F3, G2, H2, J2, and K2, Five New Oleanene-Triterpene Oligoglycosides from the Root of *Glycyrrhiza uralensis*. Chem. Pharm. Bull. (Japan), **39**, 244 (1991).

205. KITAGAWA, I., K. HORI, M. SAKAGAMI, J.-L. ZHOU, and M. YOSHIKAWA: Saponin and Sapogenol. 48. On the Constituents of the Roots of *Glycyrrhiza uralensis* Fischer from Northeastern China. 2. Licorice-Saponins D3, E2, F3, G2, H2, J2, and K2. Chem. Pharm. Bull. (Japan), **41**, 1337 (1993).

206. KITAGAWA, I., K. HORI, E. UCHIDA, W.-Z. CHEN, M. YOSHIKAWA, and J. REN: Saponin and Sapogenol. 50. On the Constituents of the Roots of *Glycyrrhiza uralensis*

Fischer from Xinjiang, China. Chemical Structures of Licorice-Saponin L3 and Isoliquiritin Apioside. Chem. Pharm. Bull. (Japan), **41**, 1567 (1993).

207. SASAKI, Y., K. MIZUTANI, R. KASAI, and O. TANAKA: Solubilizing Properties of Glycyrrhizin and Its Derivatives: Solubilization of Saikosaponin-a, the Saponin of *Bupleuri* Radix. Chem. Pharm. Bull. (Japan), **36**, 3491 (1988).

208. CAI, L.-N., R.-Y. ZHANG, Z.-L. ZHANG, B. WANG, L. QIAO, L.-R. HUANG, and J.-R. CHENG: The Structure of Glyeurysaponin. Yaoxue Xuebao (Acta Pharm. Sinica, China), **26**, 447 (1991); Chem. Abstr., **115**, 252089k (1991).

209. KITAGAWA, I., M. SAKAGAMI, F. HASHIUCHI, J.-L. ZHOU, M. YOSHIKAWA, and J.-L. REN: Apioglycyrrhizin and Araboglycyrrhizin, Two New Sweet Oleanene-Type Triterpene Oligoglycosides from the Root of *Glycyrrhiza inflata*. Chem. Pharm. Bull. (Japan), **37**, 551 (1989).

210. KITAGAWA, I., K. HORI, M. SAKAGAMI, F. HASHIUCHI, M. YOSHIKAWA, and J. REN: Saponin and Sapogenol. 49. On the Constituents of the Roots of *Glycyrrhiza inflata* Batalin from Xinjiang, China. Characterization of Two Sweet Oleanane-Type Triterpene Oligoglycosides, Apioglycyrrhizin and Araboglycyrrhizin. Chem. Pharm. Bull. (Japan), **41**, 1350 (1993).

211. ZOU, K., and R.-Y. ZHANG: Constituents from *Glycyrrhiza inflata* and Antioxidant Activities of Phenols from the Roots of *Glycyrrhiza* Spp. (Abstract of Master thesis of K. ZOU). J. Chin. Pharm. Sci., **3**, 90 (1994); Chem. Abstr., **122**, 51291r (1995).

212. ZOU, K., L.-N. CAI, and R.-Y. ZHANG: Structure Identification of Inflasaponin III and Inflasaponin V. Gaodeng Xuexiao Huaxue Xuebao (Chem. J. Chin. Univ.), **15**, 845 (1994); Chem. Abstr., **121**, 297132j (1994).

213. ZOU, K., Y.-Y. ZHAO, and R.-Y. ZHANG: Structure Determination of Two Saponins from *Glycyrrhiza inflata* Bat. Yaoxue Xuebao (Acta Pharm. Sinica, China), **29**, 393 (1994); Chem. Abstr., **121**, 251348f (1994).

214. ZOU, K., Y.-Y. ZHAO, and R.-Y. ZHANG: Structure Determination of Inflasaponin II and VI from *Glycyrrhiza inflata* Root. J. Chin. Pharm. Sci., **4**, 113 (1995); Chem. Abstr., **124**, 50610y (1996).

215. VARSHNEY, I.P., D.C. JAIN, and H.C. SRIVASTAVA: Study of Saponins from *Glycyrrhiza glabra* Root. Int. J. Crude Drug Res., **21**, 169 (1983); Chem. Abstr., **101**, 187933a (1984).

216. GAO, D.-Y., R.-Y. ZHANG, and B. YAO: Structure Identification of Yunnanglysaponin A and B. J. Chin. Pharm. Sci., **3**, 1 (1994); Chem. Abstr., **122**, 76495y (1995).

217. OHTANI, K., K. OGAWA, R. KASAI, C.-R. YANG, K. YAMASAKI, J. ZHOU, and O. TANAKA: Studies on the Constituents of *Glycyrrhiza yunnanensis*. 1. Oleanane Glycosides from *Glycyrrhiza yunnanensis* Roots. Phytochem., **31**, 1747 (1992).

218. OHTANI, K., R. KASAI, C.-R. YANG, K. YAMASAKI, J. ZHOU, and O. TANAKA: Studies on the Constituents of *Glycyrrhiza yunnanensis*. 2. Oleanane Glycosides from Roots of *Glycyrrhiza yunnanensis*. Phytochem., **36**, 139 (1994).

219. SHIBANO, M., Y. MATUMOTO, G. KUSANO, and T. SHIBATA: Researches of *Glycyrrhiza* Species Grown at Medicinal Plant Gardens in Japan and Basic Studies for Identification. 42nd Annual Meeting of Japanese Society of Pharmacognosy, Abstract Papers, p. 237. Fukuyama, September 1995.

220. HAYASHI, H., H. FUKUI, and M. TABATA: Biotransformation of 18-β-Glycyrrhetinic Acid by Cell Suspension Cultures of *Glycyrrhiza glabra*. Phytochem., **29**, 2149 (1990).

221. HAYASHI, H., K. YAMADA, H. FUKUI, and M. TABATA: Metabolism of Exogenous 18β-Glycyrrhetinic Acid in Cultured Cells of *Glycyrrhiza glabra*. Phytochem., **31**, 2729 (1992).

222a. HAYASHI, H., S. HANAOKA, S. TANAKA, H. FUKUI, and M. TABATA: Glycyrrhetinic Acid 24-Hydroxylase Activity in Microsomes of Cultured Licorice Cells. Phytochem., **34**, 1303 (1993).

222b. HAYASHI, H., Y. NISHIYAMA, N. TOMIZAWA, N. HIRAOKA, and Y. IKESHIRO: UDP-Glucurinic Acid: Triterpene Glucuronosyltransferase Activity in Cultured Licorice Cells. Phytochem., **42**, 665 (1996).

223. DORISSE, P., J. GLEYE, P. LOISEAU, P. PUIG, A.M. EDY, and M. HENRY: Papaverine Biotransformation in Plant Cell Suspension Cultures. J. Nat. Prod., **51**, 532 (1988).

224. ASADA, Y., H. SAITO, T. YOSHIKAWA, K. SAKAMOTO, and T. FURUYA: Biotransformation of 18β-Glycyrrhetinic Acid by Ginseng Hairy Root Culture. Phytochem., **34**, 1049 (1993).

225. ZENG, L., R.-Y. ZHANG, and Z.-C. LOU: Separation and Quantitative Determination of Three Saponins in Licorice Root by High-Performance Liquid Chromatography. Yaoxue Xuebao (Acta Pharm. Sinica, China), **26**, 53 (1990); Chem. Abstr., **114**, 192690s (1991).

226. KATTAEV, N.S., and G.K. NIKONOV: Glabranin, a New Flavanone from *Glycyrrhiza glabra*. Khim. Prir. Soedin., **1972**, 805; Chem. Nat. Comp. (Engl. transl.), **8**, 790 (1974).

227a. ISHII, Y., and N. SUGAWARA: Pharmacological Studies of FM 100, an Antiulcer Fraction of Licorice Root. Ouyou Yakuri (Applied Pharmacology, Japan), **7**, 871 (1973); Chem. Abstr., **80**, 66677j (1974).

227b. ITOH, M., N. ASAKAWA, Y. HASHIMOTO, M. ISHIBASHI, and H. MIYAZAKI: Quantitative Analysis of Glycyrrhizin and Glycyrrhetinic Acid in Plasma after Administration of FM-100 by Using Gas Chromatography-Selected Ion Monitoring. Yakugaku Zasshi, **105**, 1150 (1985).

228. HATANO, T., T. FUKUDA, Y.-Z. LIU, T. NORO, and T. OKUDA: Phenolic Constituents of Licorice. 4. Correlation of Phenolic Constituents and Licorice Specimens from Various Sources, and Inhibitory Effects of Licorice Extracts on Xanthine Oxidase and Monoamine Oxidase. Yakugaku Zasshi, **111**, 311 (1991).

229. SHINODA, J., and S. UEEDA: A New Flavanone Glycoside from *Glycyrrhiza glabra* L. var. *glandulifera* Regel *et* Herder. Yakugaku Zasshi, **54**, 707 (1934).

230. SHINODA, J., and S. UEEDA: The Flavanone Glycoside in *Glycyrrhiza glabra* L. var. *glandulifera* Regel *et* Herder. Chem. Ber., **67**, 434 (1934).

231. PURI, B., and T.R. SESHADRI: Survey of Anthoxanthines. 5. Coloring Matter of Liquorice Roots. J. Sci. Ind. Res., **13B**, 475 (1954).

232a. LITVINENKO, V.I., and G.V. OBOLENTSEVA: Chemical and Pharmacological Investigation of Flavonoids from Licorice (*Glycyrrhiza glabra* and *Glycyrrhiza uralensis*). Med. Prom. SSSR, **18**, 20 (1964); Chem. Abstr., **62**, 8286a (1965).

232b. ARAKAWA, H., and M. NAKAZAKI: Absolute Configuration of (–)-Hesperetin and (–)-Liquiritigenin. Chem. Ind. (London), **1960**, 73; Chem. Abstr., **54**, 9901a (1960).

233. GAFFIELD, W.: Circular Dichroism, Optical Rotatory Dispersion and Absolute Configuration of Flavanones, 3-Hydroxyflavanones and Their Glycosides; Determination of Aglycone Chirality in Flavanone Glycosides. Tetrahedron, **26**, 4093 (1970).

234. LITVINENKO, V.I., N.P. MAKSYUTINA, and I.G. KOLESNIKOV: Flavonoid Compounds of *Glycyrrhiza glabra*. Zhurn. Obshchei Khimii (USSR), **33**, 296 (1963); Chem. Abstr., **59**, 1744b (1963).

235. HOTON-DORGE, M.: Identification of Some Flavonoid Aglycone Extracts of *Glycyrrhiza glabra* Roots. J. Pharm. Belg., **29**, 560 (1974); Chem. Abstr., **83**, 111099x (1975).

236. Litvinenko, V.I.: Chalcone Glycoside of *Glycyrrhiza glabra*. Dokl. Akad. Nauk (SSSR), **155**, 600 (1964); Chem. Abstr., **60**, 14579d (1964).
237. Afchar, D., A. Cave, and J. Vaquette: Study of Licorices of Iran. 1. Flavonoids of *Glycyrrhiza glabra* var. *glandulifera*. Plant. Med. Phytother., **14**, 46 (1980); Chem. Abstr., **93**, 91865v (1980).
238. Nakanishi, T., A. Inada, K. Kambayashi, and K. Yoneda: Flavonoid Glycosides of the Roots of *Glycyrrhiza uralensis*. Phytochem., **24**, 339 (1985).
239. Miething, H., A. Speicher-Brinker, and R. Haensel: HPLC Analysis of Flavonoids from Licorice Root and their Pharmaceutical Preparations. PZ Wiss., **3**, 253 (1990); Chem. Abstr., **115**, 99427x (1991).
240. van Hulle, C., P. Braeckman, and M. Vandewalle: Isolation of Two New Flavonoids from the Root of *Glycyrrhiza glabra* var. *typica*. Planta Med., **20**, 278 (1971).
241. Litvinenko, V.I.: Flavonoid Compounds of *Glycyrrhiza uralensis*. Farmatsevt. Zh. (Kiev), **18**, 20 (1963); Chem. Abstr., **60**, 6700g (1964).
242. Litvinenko, V.I., and I.P. Kovalev: Licuraside, a New Flavonoid Glycoside of *Glycyrrhiza glabra*. Dokl. Akad. Nauk (SSSR), **169**, 347 (1966); Doklady (Engl. transl.), **169**, 691 (1966), Chem. Abstr., **65**, 13811c (1966).
243. Miething, H., and A. Speicher-Brinker: Neolicuroside, a New Chalcone Glycoside from the Roots of *Glycyrrhiza glabra*. Arch. Pharm. (Weinheim), **322**, 141 (1989); Chem. Abstr., **111**, 4253t (1989).
244. Asahina, Y.: Shosoin Yakubutu (The Shosoin Medicinals, Report on Scientific Researches), Osaka (Japan): Shokubutu Bunnkenn Kankokai, 1955.
245. Shibata, S.: Chemical Investigation of the Crude Drugs Stored in Shosoin for over Twelve Hundred Years. Int. J. Pharmacog., **32**, 75 (1994); Biol. Abstr., **97**, 160380 (1994).
246. Bhardwaj, D.K., R. Murari, T.R. Seshadri, and R. Singh: Occurrence of 2-Methylisoflavones in *Glycyrrhiza glabra*. Phytochem., **15**, 352 (1976).
247. Bhardwaj, D.K., R. Murari, T.R. Seshadri, and R. Singh: Liqcoumarin, a Novel Coumarin from *Glycyrrhiza glabra*. Phytochem., **15**, 1182 (1976).
248. Bhardwaj, D.K., and R. Singh: Glyzaglabrin, a New Isoflavone from *Glycyrrhiza glabra*. Curr. Sci. (India), **46**, 753 (1977); Chem. Abstr., **88**, 19048k (1978).
249. Bhardwaj, D.K., T.R. Seshadri, and R. Singh: Glyzarin, a New Isoflavone from *Glycyrrhiza glabra*. Phytochem., **16**, 402 (1977).
250. Saitoh, T., T. Kinoshita, and S. Shibata: Chemical Studies on the Oriental Plant Drugs. 39. New Isoflavan and Flavanone from Licorice Root. Chem. Pharm. Bull. (Japan), **24**, 752 (1976).
251. Kinoshita, T., T. Saitoh, and S. Shibata: Chemical Studies on the Oriental Plant Drugs. 40. The Occurrence of an Isoflavene and the Corresponding Isoflavone in Licorice Root. Chem. Pharm. Bull. (Japan), **24**, 991 (1976).
252. Fukai, T., B.-S. Cai, T. Horikoshi, and T. Nomura: Phenolic Constituents of *Glycyrrhiza* Species. 21. Isoprenylated Flavonoids from Underground Parts of *Glycyrrhiza glabra*. Phytochem., **43**, 1119 (1996).
253. Mitscher, L.A., Y.H. Park, S. Omoto, G.W. Clark, III, and D. Clark: Antimicrobial Agents from Higher Plants, *Glycyrrhiza glabra* L. (var. Spanish). 1. Some Antimicrobial Isoflavans, Isoflavenes, Flavanones and Isoflavones. Heterocycles, **9**, 1533 (1978).
254. Mitscher, L.A., Y.H. Park, D. Clark, and J.L. Beal: Antimicrobial Agents from Higher Plants. Antimicrobial Isoflavonoids and Related Substances from *Glycyrrhiza glabra* L. var. *typica*. J. Nat. Prod., **43**, 259 (1980).

255. KINOSHITA, T., K. KAJIYAMA, Y. HIRAGA, K. TAKAHASHI, Y. TAMURA, and K. MIZUTANI: Isoflavan Derivatives from *Glycyrrhiza glabra* (Licorice). Heterocycles, **43**, 581 (1996).

256. KAJIYAMA, K., Y. HIRAGA, K. TAKAHASHI, Y. TAMURA, K. MIZUTANI, K. OKADA, and T. KINOSHITA: Constituents of Licorice from Former USSR. 1. 114th Annual Meeting of Pharmaceutical Society of Japan, Abstract Papers, part 2, p. 204. Tokyo, March 1994.

257. KINOSHITA, T., K. KAJIYAMA, Y. HIRAGA, K. TAKAHASHI, Y. TAMURA, and K. MIZUTANI: The Isolation of new Pyrano-2-arylbenzofuran Derivatives from the Root of *Glycyrrhiza glabra*. Chem. Pharm. Bull. (Japan), **44**, 1218 (1996).

258. TAMURA, Y. K. MIZUTANI, and T. KINOSHITA: Constituents of Licorice from Former USSR. 3. 116th Annual Meeting of Pharmaceutical Society of Japan, Abstract Papers, part 2, p. 209. Kanazawa, March 1996.

259. IMOTO, K., Y. TAMURA, K. MIZUTANI, H. HARAGUTI, T. KINOSHITA: Bioactive of Phenolic Constituents of Licorice in Former USSR. 43rd Annual Meeting of Japanese Society of Pharmacognosy, Abstract Papers, p. 230. Tokyo, September 1996.

260. KITAGAWA, I., W.-Z. CHEN, K. HORI, E. HARADA (née UCHIDA), N. YASUDA, M. YOSHIKAWA, and J. REN: Chemical Studies of Chinese Licorice-Roots. 1. Elucidation of Five New Flavonoid Constituents from the Roots of *Glycyrrhiza glabra* L. Collected in Xinjiang. Chem. Pharm. Bull. (Japan), **42**, 1056 (1994).

261. FUKAI, T., J. NISHIZAWA, M. YOKOYAMA, L. TANTAI, and T. NOMURA: Phenolic Constituents of *Glycyrrhiza* Species. 16. Five New Isoprenoid-Substituted Flavonoids, Kanzonols M–P and R, from Two *Glycyrrhiza* Species. Heterocycles, **38**, 1089 (1994).

262. FUKAI, T., L. TANTAI, and T. NOMURA: NMR Spectra of Isoprenoid Substituted Phenols. 8. ^{1}H NMR Chemical Shift of the Isoflavanone 5-Hydroxyl Proton as a Characterization of 6- or 8-Prenyl Group. Heterocycles, **37**, 1819 (1994).

263. FUKAI, T., L. TANTAI, and T. NOMURA: Phenolic Constituents of *Glycyrrhiza* species. 20. Isoprenoid-Substituted Flavonoids from *Glycyrrhiza glabra*. Phytochem., **43**, 531 (1996).

264. YAHARA, S., and I. NISHIOKA: Flavonoid Glucoside from Licorice. Phytochem., **23**, 2108 (1984).

265. TAKAGI, M., T. HATANO, H. ITOH, and T. YOSHIDA: Polyphenols of Licorice. 13. 43rd Annual Meeting of Japanese Society of Pharmacognosy, Abstract Papers, p. 99. Tokyo, September 1996.

266. ZHANG, H.-J., Y. LIU, and R.-Y. ZHANG: Chemical Studies of Flavonoid Compounds from *Glycyrrhiza uralensis* Fisch. Yaoxue Xuebao (Acta. Pharm. Sinica, China), **29**, 471 (1994); Chem. Abstr., **121**, 200937q (1994).

267. SHIBATA, S., and T. SAITOH: The Chemical Studies on the Oriental Plant Drugs. 19. Some New Constituents of Licorice Root. 1. The Structure of Licoricidin. Chem. Pharm. Bull. (Japan), **16**, 1932 (1968).

268. SAITOH, T., and S. SHIBATA: Chemical Studies on the Oriental Plant Drugs. 22. Some New Constituents of Licorice Root. 2. Glycyrol, 5-*O*-Methylglycyrol and Isoglycyrol. Chem. Pharm. Bull. (Japan), **17**, 729 (1969).

269. KANEDA, M., T. SAITOH, Y. IITAKA, and S. SHIBATA: Chemical Studies on the Oriental Plant Drugs. 36. Structure of Licoricone, a New Isoflavone from Licorice Root. Chem. Pharm. Bull. (Japan), **21**, 1338 (1973).

270. KANEDA, M., Y. IITAKA, and S. SHIBATA: The Crystal and Molecular Structure of Licoricone Monobromoacetate. Acta. Crystallogr., **B29**, 2827 (1973).

271. TSUKAYAMA, M., T. HORIE, K. FUJIMOTO, and M. NAKAYAMA: A Convenient Synthesis of Prenylated Isoflavones: Synthesis of Licoricone and Related Compounds. Chem. Pharm. Bull. (Japan), **34**, 2369 (1986).

272. FUKAI, T., M. TOYONO, and T. NOMURA: Phenolic Constituents of *Glycyrrhiza* Species. 1. On the Structure of Licoricidin. Heterocycles, **27**, 2309 (1988).

273. LAM, Y.K.T., M. SANDRINO-MEINZ, L. HUANG, R.D. BUSCH, T. MELLIN, D. ZINK, and G.Q. HAN: 5-*O*-Methyllicoricidin: a New and Potent Benzodiazepine-Binding Stimulator from *Glycyrrhiza uralensis*. Planta Med., **58**, 221 (1992).

274a. SHIZURI, Y., K. UCHIDA, and S. YAMAMURA: Isolation of Licorisoflavans A and B, Muscle Relaxant Substances in Licorice. Abstract Paper of 56th Annual Meeting of Jpn. Chem. Soc., p. 1154. Tokyo, April 1988.

274b. KAWAKAMI, T., M. UCHIDA, M. WADA, T. YAMAMURA, and F. HARADA: Novel Isoflavan Derivatives, and Their Preparation from Licorice Roots and Use as Muscle Relaxants. Jpn. patent; Chem. Abstr., **111**, 167414k (1988).

275. SHIH, T.L., M.J. WYVRATT, and H. MROZIK: Total Synthesis of (±)-5-*O*-Methyllicoricidin. J. Organ. Chem. (USA), **52**, 2029 (1987).

276. FUKAI, T., and T. NOMURA: NMR Spectra of Isoprenoid Substituted Phenols. 5. ^{1}H-NMR Chemical Shift of the Flavonol 5-Hydroxy Proton as a Characterization of 6- or 8-Isoprenoid Substitution. Heterocycles, **34**, 1213 (1992).

277. FUKAI, T., Q.-H. WANG, T. KITAGAWA, K. KUSANO, T. NOMURA, and Y. IITAKA: Phenolic Constituents of *Glycyrrhiza* Species. 3. Structures of Six Isoprenoid-Substituted Flavonoids, Gancaonins F, G, H, I, Glycyrol, and Isoglycyrol from Xibei Licorice (*Glycyrrhiza* Spp.). Heterocycles, **29**, 1761 (1989).

278. SHIOZAWA, T., S. URATA, T. KINOSHITA, and T. SAITOH: Revised Structures of Glycyrol and Isoglycyrol, Constituents of the Root of *Glycyrrhiza uralensis*: Chem. Pharm. Bull. (Japan), **37**, 2239 (1989).

279. CHANG, X.-R., Q.-H. XU, D.-Y. ZHU, G.-Q. SONG, and R.-S. XU: Constituents of Licorice. 1. Antimicrobial Active Compound, Structure of Licobenzofuran (Abstract of 1st Japan-China Congress of Chemistry of Natural Products, Nagoya (Japan), 1981. Zhongcaoyao (Chin. Herbal Med.), **12**, 530 (1981); Chem. Abstr., **97**, 20701k (1982).

280. CHANG, X.-R., Q.-H. XU, D.-Y. ZHU, G.-Q. SONG, and R.-S. XU: The Isolation and Structural Elucidation of Liconeolignan from *Glycyrrhiza uralensis*. Yaoxue Xuebao (Acta Pharm. Sinica, China), **18**, 45 (1983); Chem. Abstr., **99**, 67496d (1983).

281. ZHU, D.-Y., G.-Q. SONG, F.-X. JIAN, X.-R. CHANG, and W.-B. GUO: Studies on Chemical Constituents of *Glycyrrhiza uralensis* Fisch. The Structures of Isolicoflavonol and Glycycoumarin. Huaxue Xuebao (Acta. Chem. Sinica, China), **42**, 1080 (1984); Chem. Abstr., **102**, 75705s (1985).

282. KIUCHI, F., X. CHEN, and Y. TSUDA: Four New Phenolic Constituents from Licorice (Root of *Glycyrrhiza* sp.). Heterocycles, **31**, 629 (1990).

283. KINOSHITA, T., T. SAITOH, and S. SHIBATA: Chemical Studies on the Oriental Drugs. 42. A New 3-Arylcoumarin from Licorice Root. Chem. Pharm. Bull. (Japan), **26**, 135 (1978).

284. FUKAI, T., J. NISHIZAWA, M. YOKOYAMA, and T. NOMURA: Phenolic Constituents of *Glycyrrhiza* Species. 13. Five New Isoprenoid-Substituted Flavonoids, Kanzonols F–J, from *Glycyrrhiza uralensis*. Heterocycles, **36**, 2565 (1993).

285. FUKAI, T., J. NISHIZAWA, and T. NOMURA: Phenolic Constituents of *Glycyrrhiza* Species. 14. Variations in the Chemical Shift of the 5-Hydroxyl Proton of Isoflavones; Two Isoflavones from Licorice. Phytochem., **36**, 225 (1994).

286. TAKAGI, M, M. MIZUNO, T. HATANO, and T. YOSHIDA: Polyphenols of Licorice. 11. 42nd Annual Meeting of Japanese Society of Pharmacognosy, Abstract Papers, p. 206. Fukuyama, September 1995.

287. SAITOH, T., H. NOGUCHI, and S. SHIBATA: Chemical Studies on the Oriental Plant Drugs. 44. A New Isoflavone and the Corresponding Isoflavanone of Licorice Root. Chem. Pharm. Bull. (Japan), **26**, 144 (1978).

288. ZENG, L., T. FUKAI, T. NOMURA, R.-Y. ZHANG, and Z.-C. LOU: Phenolic Constituents of *Glycyrrhiza* Species. 8. Four New Prenylated Flavonoids, Glyasperins A, B, C, and D from the Roots of *Glycyrrhiza aspera*. Heterocycles, **34**, 575 (1992).

289. ZENG, L., T. FUKAI, T. NOMURA, R.-Y. ZHANG, and Z.-C. LOU: Phenolic Constituents of *Glycyrrhiza* Species. 10. Glyasperin E, a New 3-Phenoxychromen-4-one Derivative from the Roots of *Glycyrrhiza aspera*. J. Chem. Soc., Perkin Trans. I. (London), **1993**, 1153.

290. ZENG, L., T. FUKAI, T. NOMURA, R.-Y. ZHANG, and Z.-C. LOU: Phenolic Constituents of *Glycyrrhiza* Species. 9. Five New Isoprenoid-Substituted Flavonoids, Glyasperins F, G, H, I, and J from the Roots of *Glycyrrhiza aspera*. Heterocycles, **34**, 1813 (1992).

291. FUKAI, T., L. ZENG, J. NISHIZAWA, Y.-H. WANG, and T. NOMURA: Phenolic Constituents of *Glycyrrhiza* Species. 15. Four Isoprenoid-Substituted Flavonoids from *Glycyrrhiza aspera*. Phytochem., **36**, 233 (1994).

292. VINCE, S., and A. KISS: Synthesis of C-3-Substituted Chromones. 6. Preparation of 2-Phenoxy-2'-hydroxyacetophenones and Cyclization to 3-Phenoxychromones. Magy. Kem. Foly. (Hungary) **85**, 353 (1979); Chem. Abstr., **92**, 110775p (1980).

293. KITAGAWA, K., W.-Z. CHEN, and K. HORI: Chemical Studies on Crude Drug Processing: Licorice. 9. Flavonoids from *Glycyrrhiza aspera* collected in Xinjiang, China. 40th Annual Meeting of the Japanese Society of Pharmacognosy, Abstract Papers, p. 96. Osaka, September 1993.

294. FUKAI, T., J. NISHIZAWA, and T. NOMURA: Phenolic Constituents of *Glycyrrhiza* Species. 12. Five Isoprenoid-Substituted Flavonoids from *Glycyrrhiza eurycarpa*. Phytochem., **35**, 515 (1994).

295. ZENG, L., T. FUKAI, T. KANEKI, T. NOMURA, R.-Y. ZHANG, and Z.-C. LOU: Phenolic Constituents of *Glycyrrhiza* Species. 7. Four New Isoprenoid-Substituted Dibenzoylmethane Derivatives, Glyinflanins A, B, C, and D from the Roots of *Glycyrrhiza inflata*. Heterocycles, **34**, 85 (1992).

296. SAITOH, T., and S. SHIBATA: New Type Chalcone from Licorice Root. Tetrahedron Letters, **1975**, 4461.

297. SAITOH, T., S. SHIBATA, U. SANKAWA, T. FURUYA, and S. AYABE: Biosynthesis of Echinatin. A New Biosynthetical Scheme of Retrochalcone. Tetrahedron Letters, **1975**, 4463.

298. XU, R.-S., K.-L. WEN, S.-F. JIANG, C.-G. WANG, F.-X. JIANG, Y.-Y. XIE, and Y.-S. GAO: The Isolation, Structure and Total Synthesis of Licochalcone. Hua Hsueh Hsueh Pro (Acta Chim. Sinica, China), **37**, 289 (1978); Chem. Abstr., **92**, 198060k (1980).

299. FUKAI, T., and T. NOMURA: Phenolic Constituents of *Glycyrrhiza* Species. 18. Isoprenoid-Substituted Flavonoids from Roots of *Glycyrrhiza inflata*. Phytochem., **38**, 759 (1995).

300. GOWAN, J.E., and T.S. WHEELER: Further Experiments on the Mechanism of the Baker-Venkataraman Transformation. J. Chem. Soc. (London), **1950**, 1925.

301. SESHADRI, T.R.: Recent Developments in the Chemistry of Flavonoids. Tetrahedron, **6**, 169 (1959).

302. Kajiyama, K., S. Demizu, Y. Hiraga, K. Kinoshita, K. Koyama, K, Takahashi, Y. Tamura, K. Okada, and T. Kinoshita: New Prenylflavones and Dibenzoylmethane from *Glycyrrhiza inflata*. J. Nat. Prod., **55**, 1197 (1992).

303. Demizu, S., K. Kajiyama, Y. Hiraga, K. Kinoshita, K. Koyama, K. Takahashi, Y. Tamura, K. Okada, and T. Kinoshita: Prenylated Dibenzoylmethane Derivatives from the Root of *Glycyrrhiza inflata* (Xinjiang Licorice). Chem. Pharm. Bull. (Japan), **40**, 392 (1992).

304. Kajiyama, K., S. Demizu, Y. Hiraga, K. Kinoshita, K. Koyama, K. Takahashi, Y. Tamura, K. Okada, and T. Kinoshita: Two Prenylated Retrochalcones from *Glycyrrhiza inflata*. Phytochem., **31**, 3229 (1992).

305. Christensen, S.B., C. Ming, L. Andersen, U. Hjørne, C.E. Olsen, C. Cornett, T.G. Theander, and A. Kharazmi: An Antileishmanial Chalcone from Chinese Licorice Roots. Planta Med., **60**, 121 (1994).

306. Zou, K., R.-Y. Zhang, and X.-B. Yang: Structure Determination of Inflacoumarin A from *Glycyrrhiza inflata*. Yaoxue Xuebao (Acta Pharm. Sinica, China), **29**, 397 (1994); Chem. Abstr., **121**, 226362y (1994).

307. Yang, Y.-B., K. Zou, B. Yao, R.-Y. Zhang: NMR Study of Inflacoumarin A. Bopuxue Zazhi (Chin. J. Mag. Res.), **11**, 399 (1994); Chem. Abstr., **122**, 101601q (1995).

308. Liu, Q., and Y.-L. Liu: Studies on Chemical Constituents of *Glycyrrhiza eurycarpa* P. C. Li. Yaoxue Xuebao (Acta Pharm. Sinica, China), **24**, 525 (1989); Chem. Abstr., **112**, 115711f (1990).

309. Liu, Q., and Y.-L. Liu: Application of ^{13}C NMR to Structural Identification of the Flavonoid Glycoside. Zhonggno Yixue Kexyeyuan Xuebao (Acta Academica Med. Sinica, China), **12**, 359 (1990); Chem. Abstr., **115**, 178517e (1991).

310. Liu, M., Q. Lin, Y.-L. Liu, C.-Y. Hou, and T.J. Mabry: An Acylated Flavone *C*-Glycoside from *Glycyrrhiza eurycarpa*. Phytochem., **36**, 1089 (1994).

311. Zhang, Y.-M., X.-D. Xu, B.-H. Hu, Q. Liu, C.-Y. Hou, Y.-L. Liu, and J.-S. Yang: Eurycarpin A, a New Isoflavone from *Glycyrrhiza eurycarpa*. Chin. Chem. Letters, **6**, 477 (1995); Chem. Abstr., **123**, 138814h (1995).

312. Fukai, T., Q.-H. Wang, R. Inami, and T. Nomura: Phenolic Constituents of *Glycyrrhiza* Species. 5. Structures of Prenylated Dihydrochalcone, Gancaonin J and Homoisoflavanone, Gancaonin K from *Glycyrrhiza pallidiflora*. Heterocycles, **31**, 643 (1990).

313. Dewick, P.M.: Biosynthesis of the 3-Benzylchroman-4-one Eucomin in *Eucomis bicolor*. Phytochem., **14**, 983 (1975).

314. Ayabe, S., T. Yoshikawa, M. Kobayashi, and T. Furuya: Studies on Plant Tissue Culture. 34. Biosynthesis of a Retrochalcone, Echinatin: Involvement of *O*-Methyltransferase to Licodione. Phytochem., **19**, 2331 (1980).

315. Kajiyama, K., Y. Hiraga, K. Takahashi, S. Hirata, S. Kobayashi, U. Sankawa, and T. Kinoshita: Flavonoids and Isoflavonoids of Chemotaxonomic Significance from *Glycyrrhiza pallidiflora* (Leguminosae). Biochem. Syst. Ecol., **21**, 785 (1993); Chem. Abstr., **120**, 212563g (1994).

316. Cai, L.-N., R.-Y. Zhang, B. Wang, L. Qiao, L.-R. Huang, and Z.-L. Zhang: Chemical Constituents of *Glycyrrhiza pallidiflora* Maxim. Yaoxue Xuebao (Acta Pharm. Sinica, China), **27**, 748 (1992); Chem. Abstr., **118**, 165182d (1993).

317. Motegi, J.: Licorice on Japanese Market, In: Records of Lecture in 2nd Meeting on Crude Drugs; Licorice, p. 31. Tokyo: Tokyo Soc. Pharmacognosy, 1986.

318. Mitsuhashi, H. (ed.): Shoyakugaku (Pharmacognosy), p. 141, Tokyo: Nankoudo, 1983.

319. NIHON YAKUKYOKUHO KAISETUSHO HENSHU IINKAI (Editorial Board of the Manual of Japanese Pharmacopoeia): Dai 13-Kaisei Nihon Yatkyokuho Kaisetusho (Manual of Japanese Pharmacopeia; 13th edn.), p. D-227, Tokyo: Hirokawa, 1996.
320. NAMBA, T.: Genshoku Hyatuka, Sekai No Yakuyou Shokubutu (The Encyclopedia of Herbs and Herbalism; original book edited by STUART, M., London: Orbis, 1979, Notes of Traditional Shino-Japanese Medicine were added by NAMBA, T.), Vol. 2, p. 212, Tokyo: Enterprise, 1988.
321. KITAGAWA, K., U. SANKAWA, J. SHOUJI, M. TAKIDO, M. TOMODA, and I. NISHIOKA: Shoyaku Gaku (Pharmacognosy), p. 218, Tokyo: Hirokawa, 1980.
322. KARIYONE, T.: Wakan Shoyaku (The Traditional Sino-Japanese Crude Drugs). p. 167, Tokyo: Hirokawa, 1971.
323. SUGAYA, I., and N. SAHASHI (eds.): Yakuyou Shokubutu Gaku (Medicinal Plants), p. 203, Tokyo: Hirokawa, 1990.
324. SATO, J.: Kanyaku No Gen-shokubutu (On Chinese Medical Plants). p. 58, Tokyo: Nihon Gakujutu Sinko-kai, 1959.
325. NAMBA, T.: Genshoku Wakanyaku Zukan (Colored Illustrations of Wakan-Yaku; The Crude Drugs in Japan, China and the neighboring Countries), vol. 1, p. 43, Osaka: Hoikusha, 1980.
326. YONEDA, K., E. YAMAGATA, and M. TSUJIMURA: Studies on Resources of Crude Drugs. 5. Comparison of the Constituents of Wild *Glycyrrhiza uralensis* and Various Chinese Licorices Obtained in Japanese Market. Shoyakugaku Zasshi (Jpn. J. Pharmacognosy), **44**, 202 (1990); Chem. Abstr., **114**, 128867c (1991).
327. YONEDA, K., E. YAMAGATA, and M. TERUYA: Studies on Resources of Crude Drugs. 6. Comparison of the Constituents of *Glycyrrhiza* Radix (Licorice) from Various Countries. 2. Shoyakugaku Zasshi (Jpn. J. Pharmacognosy), **45**, 220 (1991); Chem. Abstr., **116**, 158674w (1992).
328. AMAGAYA, S., E. SUGISHITA, Y. OGIHARA, S. OGAWA, K. OKADA, and T. AIZAWA: Separation and Quantitative Analysis of 18α-Glycyrrhetinic Acid and 18β-Glycyrrhetinic Acid in Glycyrrhizae Radix by Gas-Liquid Chromatography. J. Chromatogr., **320**, 430 (1985).
329. HIRAGA, Y., H. ENDO (née KAIZUKA), K. TAKAHASHI, and S. SHIBATA: High-Performance Liquid Chromatographic Analysis of Licorice Extracts. J. Chromatogr., **292**, 451 (1984).
330. HATTORI, M., K. MIYACHI, Y.-Z. SHU, N, KAKIUCHI, and T. NAMBA: Studies on Dental Caries Prevention by Traditional Medicines. 9. Potent Antibacterial Action of Coumarin Derivatives from Licorice Roots against *Streptococcus mutants*. Shoyakugaku Zasshi (Jpn. J. Pharmacognosy), **40**, 406 (1986); Chem. Abstr., **107**, 46132a (1987).
331. SHIBATA, S., and T. HIRAGA: Analysis of Herb Medicines by Three-dimensional High-Performance Liquid Chromatography. Analysis of *Glycyrrhiza* Components. Pharm. Tech. Jpn., **2**, 569 (1986); Chem. Abstr., **106**, 125947c (1987).
332. DEMIZU, S., K. KAJIYAMA, K. TAKAHASHI, Y. HIRAGA, S. YAMAMOTO, Y. TAMURA, K. OKADA, and T. KINOSHITA: Antioxidant and Antimicrobial Constituents of Licorice: Isolation and Structure Elucidation of a New Benzofuran Derivative. Chem. Pharm. Bull. (Japan), **36**, 3474 (1988).
333. SAITOH, T., T. KINOSHITA, and S. SHIBATA: Chemical Studies on the Oriental Plant Drugs, 41. Flavonols of Licorice Root. Chem. Pharm. Bull. (Japan), **24**, 1242 (1976).
334. KINOSHITA, T., T. SAITOH, and S. SHIBATA: Chemical Studies on the Oriental Plant Drugs. 43. A New Isoflavone from Licorice Root. Chem. Pharm. Bull. (Japan), **26**, 141 (1978).

335. Hatano, T., H. Kagawa, T. Yasuhara, and T. Okuda: Two New Flavonoids and Other Constituents in Licorice Root: Their Relative Astringency and Radical Scavenging Effects. Chem. Pharm. Bull. (Japan), **36**, 2090 (1988).

336. Aga, Y., K. Tutumi, T. Hatano, and T. Okuda. Polyphenols of Licorice. 9. 39th Annual Meeting of the Japanese Society of Pharmacognosy, Abstract Papers, p. 141. Tokyo, September 1992.

337. Hatano. T., T. Yasuhara, T. Fukuda, T. Noro, and T. Okuda: Phenolic Constituents of Licorice. 2. Structure of Licopyranocoumarin, Licoarylcoumarin and Glisoflavone, and Inhibitory Effects of Licorice Phenolics on Xanthine Oxidase. Chem. Pharm. Bull. (Japan), **37**, 3005 (1989).

338. Hatano, T., T. Fukuda, T. Miyase, T. Noro, and T. Okuda: Phenolic Constituents of Licorice. 3. Structures of Glicoricone and Licofuranone, and Inhibitory Effects of Licorice Constituents on Monoamine Oxidase. Chem. Pharm. Bull. (Japan), **39**, 1238 (1991).

339. Fukai, T., H. Kato, and T. Nomura: Phenolic Constituents of *Glycyrrhiza* Species. 11. A New Prenylated 3-Arylcoumarin, Gancaonin W, from Licorice. Shoyakugaku Zasshi (Jpn. J. Pharmacognosy), **47**, 326 (1993); Chem. Abstr., **120**, 294089u (1994).

340. Fukai, T., Y.-H. Wang, J. Nishizawa, S.-R. Xing, and T. Nomura: Phenolic Constituents of *Glycyrrhiza* Species. 17. Three Isoprenoid-Substituted Isoflavans, Gancaonins X - Z, from Chinese Folk Medicine "Tiexin Gancao" (Root Xylems of *Glycyrrhiza* Species). Nat. Med. (Jpn. J. Pharmacognosy), **48**, 203 (1994); Chem. Abstr., **122**, 114727k (1995).

341. Fukui, H., K. Goto, and M. Tabata: Two Antimicrobial Flavanones from the Leaves of *Glycyrrhiza glabra*. Chem. Pharm. Bull. (Japan), **36**, 4174 (1988).

342. Litvinenko, V.I., and T.P. Nadezhina: Flavonoids from the Aboveground Part of *Glycyrrhiza glabra*. Rast. Resur., **8**, 35 (1972); Chem. Abstr., **76**, 124123f (1972).

343. Bohm, B.A.: Flavanones and Dihydroflavonols. In: The Flavonoids, 1 (Harborne, J.B., T.J. Mabry, and H. Mabry, eds.), p. 597, London: Chapman and Hall, 1975.

344. Litvinenko, V.I.: Glyphoside, a New Flavonoid Glycoside obtained from *Glycyrrhiza glabra*. Rast. Resur., **2**, 531 (1966); Chem. Abstr., **66**, 73264m (1967).

345. Ingham, J.L.: An Isoflavan Phytoalexin from Leaves of *Glycyrrhiza glabra*. Phytochem., **16**, 1457 (1977).

346. Fukai, T., Q.-H. Wang, and T. Nomura: Phenolic Constituents of *Glycyrrhiza* Species. 2. Four New Prenylated Flavonoids from Aerial Parts of *Glycyrrhiza uralensis*. Heterocycles, **29**, 1369 (1989).

347. Fukai, T., Q.-H. Wang, M. Takayama, and T. Nomura: Phenolic Constituents of *Glycyrrhiza* Species. 4. Structures of Five New Prenylated Flavonoids, Gancaonins L, M, N, O, and P from Aerial Parts of *Glycyrrhiza uralensis*. Heterocycles, **31**, 373 (1990).

348. Fukai, T., Q.-H. Wang, and T. Nomura: Phenolic Constituents of *Glycyrrhiza* Species. 6. Six Prenylated Phenols from *Glycyrrhiza uralensis*. Phytochem., **30**, 1245 (1991).

349. Takayama, M., T. Fukai, K. Ichikawa, and T. Nomura: Identification of Prenylated Flavonoids Using Fast-Atom Bombardment Mass Spectrometry. Rapid Commun. Mass Spectro., **5**, 67 (1991).

350. Takayama, M., T. Fukai, Y. Hano, and T. Nomura: Mass Spectrometry of Prenylated Flavonoids. Heterocycles, **33**, 405 (1992).

351. Breitmaier, E., and W. Voelter: Carbon-13 NMR Spectroscopy, 3rd edn., Chapter 5, Weinheim: VCH, 1987.
352. Fukai, T., and T. Nomura: Variations in the Chemical Shift of Benzylic Methylene Carbon of Prenyl Group on Heterocyclic Prenylphenols. Heterocycles, **42**, 911 (1996).
353. Jia, S.-S., C.-M. Ma, and J.-M. Wang: Studies of Flavonoid Constituents Isolated from the Leaves of *Glycyrrhiza uralensis* Fisch. Yaoxue Xuebao (Acta Pharm. Sinica, China), **25**, 758 (1990); Chem. Abstr., **117**, 86658q (1992).
354. Jia, S.-S., D. Liu, X.-P. Zheng, Y. Zhang, and Y.-K. Li: Two New Isoprenyl Flavonoids from the Leaves of *Glycyrrhiza uralensis* Fisch. Chin. Chem. Letters, **3**, 189 (1992); Chem. Abstr., **121**, 226357a (1994).
355. Jia, S.-S., D. Liu, X.-P. Zheng, Y. Zhang, and Y.-K. Li: Two New Isoprenyl Flavonoids from the Leaves of *Glycyrrhiza uralensis* Fisch. Yaoxue Xuebao (Acta Pharm. Sinica, China), **28**, 28 (1993); Chem. Abstr., **119**, 221612d (1993).
356. Jia, S.-S., D. Liu, H.-Q. Wang, and Z.-X. Suo: Isolation and Identification of Gancaonin P 3′-Methyl Ether from the Leaves of *Glycyrrhiza uralensis* Fisch. Yaoxue Xuebao (Acta Pharm. Sinica, China), **28**, 623 (1993); Chem. Abstr., **120**, 4628e (1994).
357. Jia, S.-S., C.-M. Ma, Y.-H. Li, and J.-H. Hao: Glycosides of Phenolic Acid and Flavonoids from the Leaves of *Glycyrrhiza uralensis* Fisch. Yaoxue Xuebao (Acta Pharm. Sinica, China), **27**, 441 (1992); Chem. Abstr., **117**, 178168h (1992).
358. Ghisalberti, E.L., P.R. Jefferies, and D. McAdam: Isoprenylated Resorcinol Derivatives from *Glycyrrhiza acanthocarpa*. Phytochem., **20**, 1959 (1981).
359. Fukai, T., L. Zeng, T. Nomura, R.-Y. Zhang, and Z.-C. Lou: Phenolic Constituents of *Glycyrrhiza* Species. 19. Phenolic Constituents of Aerial Parts of *Glycyrrhiza eurycarpa*. Nat. Med. (Jpn. J. Pharmacognosy), **50**, 274 (1996); Chem. Abstr., **125**, 163282g (1996).
360. Mitscher, L.A., G.S. Raghav Rao, I. Khanna, T. Veysoglu, and S. Drake: Antimicrobial Agents from Higher Plants: Prenylated Flavonoids and Other Phenols from *Glycyrrhiza lepidota*. Phytochem., **22**, 573 (1983).
361. Gollapudi, S.R., H. Telikepalli, A. Keshavarz-Shokri, D.V. Velde, and L.A. Mitscher: Glepidotin C: a Minor Antimicrobial Bibenzyl from *Glycyrrhiza lepidota*. Phytochem., **28**, 3556 (1989).
362. Jain, A.C., and M.K. Zutshi: The Synthesis of Sericetin and Related Flavonols. Tetrahedron, **29**, 3347 (1973).
363. Furuya. T., K. Matsumoto, and M. Hikichi: Studies on Plant Tissue Cultures. 11. Echinatin, a New Chalcone from Tissue Culture of *Glycyrrhiza echinata*. Tetrahedron Letters, **1971**, 2567.
364. Furuya, T., S. Ayabe, and M. Kobayashi: Studies on Plant Tissue Cultures. 29. Licodione, a New Dibenzoylmethane Derivative from Cultured Cells of *Glycyrrhiza echinata*. Tetrahedron Letters, **1976**, 2539.
365. Ayabe, S., and T. Furuya: ^{13}C NMR Studies on Licodione and Related Compounds in Equilibrium Mixture of Keto and Enol Forms. Tetrahedron Letters, **21**, 2965 (1980).
366. Ayabe, S., K. Iida, and T. Furuya: Studies on Plant Tissue Cultures. 46. Stress-Induced Formation of Echinatin and a Metabolite, 5′-Prenyl-licodione, in Cultured *Glycyrrhiza echinata* Cells. Phytochem., **25**, 2803 (1986).
367. Ayabe, S., M. Kobayashi, M. Hikichi, K. Matsumoto, and T. Furuya: Studies on Plant Tissue Cultures. 33. Flavonoids from the Cultured Cells on *Glycyrrhiza echinata*. Phytochem., **19**, 2179 (1980).

368. AYABE, S., and T. FURUYA: Studies on Plant Tissue Cultures. 35. Biosynthesis of a Retrochalcone, Echinatin: a Feeding Study with Advanced Precursors. Tetrahedron Letters, **22**, 2097 (1981).

369. AYABE, S., and T. FURUYA: Studies on Plant Tissue Cultures. 36. Biosynthesis of a Retrochalcone, Echinatin, and Other Flavonoids in the Cultured Cells of *Glycyrrhiza echinata*. A New Route to a Chalcone with Transposed A- and B-Rings. J. Chem. Soc. Perkin Trans. I (London), **1982**, 2725.

370. AYABE, S., T. YOSHIKAWA, and T. FURUYA: Biosynthesis of Retrochalcone and Related Flavonoids; the Tissue Culture of *Glycyrrhiza echinata*, In: Plant Tissue Culture 1982. Proc. 5th Intl. Cong. Plant Tissue and Cell Culture. (FUJIWARA, A, ed.), pp. 379–380. Tokyo: Maruzen, 1982; Chem. Abstr., **99**, 155308z (1983).

371. AYABE, S., K. IIDA, and T. FURUYA: Studies on Plant Tissue Cultures. 45. Induction of Stress Metabolites in Immobilized *Glycyrrhiza echinata* Cultured Cells. Plant Cell. Rep., **5**, 186 (1986).

372. AYABE, S., A. UDAGAWA, K. IIDA, T. YOSHIKAWA, and T. FURUYA: Studies on Plant Tissue Cultures. 47. Regulation of Retrochalcone Biosynthesis: Activity Changes of *O*-Methyltransferases in the Yeast Extract-Induced *Glycyrrhiza echinata* Cells. Plant Cell Rep., **6**, 16 (1987).

373. AYABE, S., A. UDAGAWA, and T. FURUYA: Studies on Plant Tissue Cultures. 54. NAD(P)H-Dependent 6′-Deoxychalcone Synthase Activity in *Glycyrrhiza echinata* Cells Induced by Yeast Extract. Arch. Biochem. Biophys., **261**, 458 (1988).

374. AYABE. S., A. UDAGAWA, and T. FURUYA: Studies on Plant Tissue Cultures. 52. Stimulation of Chalcone Synthase Activity by Yeast Extract in Cultured *Glycyrrhiza echinata* Cells and 5-Deoxyflavanone Formation by Isolated Protoplasts. Plant Cell Rep., **7**, 35 (1988).

375. HARANÔ, K., N. OKADA, T. FURUNO, T. TAKAHASHI, S. AYABE, and R. WELLE: Enzymatic Synthesis of 6′-Deoxychalcone in Cultured *Glycyrrhiza echinata* Cells. Plant Cell Rep., **12**, 66 (1993).

376. OTANI, K., T. TAKAHASHI, T. FURUYA, and S. AYABE: Licodione Synthase, a Cytochrome P450 Monooxygenase Catalyzing 2-Hydroxylation of 5-Deoxyflavanone, in Cultured *Glycyrrhiza echinata* L. Cells. Plant Physiol., **105**, 1427 (1994).

377. KOBAYASHI, A., S. YATA, and K. KAWAZU: A β-Hydroxychalcone and Flavonoids from Alfalfa Callus Stimulated by a Fungal Naphthoquinone, PO-1. Agric. Biol. Chem., **52**, 3223 (1988).

378. KIRIKAE, Y., M. SAKURAI, T. FURUNO, T. TAKAHASHI, and S. AYABE: Biosynthesis of a Dibenzoylmethane, Licodione, in Cultured Alfalfa Cells Induced by Yeast Extract. Biosci. Biotech. Biochem., **57**, 1353 (1993).

379. ZENG, L., R.-Y. ZHANG, T. MENG, and Z.-C. LOU: Determination of Nine Flavonoids and Coumarins in Licorice Root by High-Performance Liquid Chromatography. J. Chromatogr., **513**, 247 (1990).

380. YANG, L., Y.-L. LIU, and S.-Q. LIN: HPLC Analysis of Flavonoids in the Root of Six *Glycyrrhiza* Species. Yaoxue Xuebao (Acta Pharm. Sinica, China), **25**, 840 (1990); Biol. Abstr., **91**, 100324 (1991).

381. KINOSHITA, T., U. SANKAWA, T. TAKUMA, K. ASAHI, and N. TAKAHASHI: Induction of Differentiation in Murine Erythroleukemia Cells by Flavonoids. Chem. Pharm. Bull. (Japan), **33**, 4109 (1985).

382. FUKAI, T., and T. NOMURA: NMR Spectra of Isoprenoid Substituted Phenols. 1. Constituents of the Moraceae Plants. 5. Revised Structures of Broussoflavonols C and D, and the Structure of Broussoflavonol E. Heterocycles, **29**, 2379 (1989).

383. NOMURA, T.: Phenolic Compounds of the Mulberry Tree and Related Plants. In: Progress Chem. Org. Nat. Prod. (Fortschr. Chem. organ. Naturstoffe), Vol. 53 (HERZ, W., H. GRISEBACH, G. W. KIRBY, and CH. TAMM, eds.), p. 87. Wien: Springer-Verlag, 1988.
384. NOMURA, T., and Y. HANO: Isoprenoid-substituted Phenolic Compounds of Moraceous Plants. Nat. Prod. Reports, **11**, 205 (1994).
385. NOMURA, T., Y. HANO, and M. AIDA: Chemistry and Biological Activity of Isoprenoid-Substituted Phenolic Compounds from *Artocarpus* Plants (Moraceae). NRCT-JSPS Core University System, Pharmaceutical Sciences and Chemistry of Natural Products and Polymer Sciences, The 3rd Joint-Seminar "Current Advances in Natural Product Research", Abstract Papers, p. 145. Bangkok, November 1996.
386. KATO, S., Y. HANO, T. FUKAI, Y. KOSUGE, and T. NOMURA: Kazinol P, a Novel Isoprenylated Spiro-Compound from *Broussonetia kazinoki* Sieb. Heterocycles, **24**, 2141 (1986).
387. FUKAI, T., and T. NOMURA: Variations in the Chemical Shift of Methylene Carbon of Prenyl Group on Naturally Occurring Phenols. 1995 International Chemical Congress of Pacific Basin Societies, Abstract Papers, part 2, p. 9-672. Honolulu, December 1995.
388. DHAMI, K.S., and J.B. STOTHERS: ^{13}C NMR Studies. 8. ^{13}C spectra of Some Substituted Anisoles, Can. J. Chem., **44**, 2855 (1966).
389. MARKHAM, K.R., and V.M. CHARI: Carbon-13 NMR Spectroscopy of Flavonoids. In: The Flavonoids: Advances in Research. (HARBORNE, J.B., and T.J. MABRY, eds.), pp. 19-51, London: Chapman and Hall, 1982.
390. TAHARA, S., M. MORIYAMA, J.L. INGHAM, and J. MIZUTANI: Structure Revision of Piscidone, a Major Isoflavonoid in the Root Bark of *Piscidia erythrina*. Phytochem., **31**, 679 (1992).
391. LIN, C.-H., P.-N. CHIU, S.-C. FANG, B.-J. SHIEH, and R.-R. WU: Revised Structure of Broussoflavonol G and the 2D NMR Spectra of Some Related Prenylflavonoids. Phytochem., **41**, 1215 (1996).
392. TAHARA, S.: Structural Diversity in Isoflavonoids and Erroneously Proposed Structures. Kagaku to Seibutsu (Chem. Biol.; J. Jpn. Agrical. Chem. Soc.), **29**, 493 (1991); Chem. Abstr., **115**, 203246a (1991).
393. FURUKAWA, H., M. YOGO, and T.-S. WU: Acridone Alkaloids. 10. ^{13}C-Nuclear Magnetic Resonance Spectra of Acridone Alkaloids. Chem. Pharm. Bull. (Japan), **31**, 3084 (1983).
394. HARBORNE, J.B., and T.J. MABRY (eds.): The Flavonoids: Advances in Research. London: Chapman and Hall, 1982.
395. HARBORNE, J.B. (ed.): The Flavonoids: Advances in Research Since 1980. London, Chapman and Hall, 1988.
396. HARBORNE, J.B. (ed.): The Flavonoids: Advances in Research Since 1986. London: Chapman and Hall, 1994.
397. BARRON, D., and R.K. IBRAHIM: Isoprenylated Flavonoids: A Survey. Phytochem., **43**, 921 (1996).
398. FEIGL, F., and V. ANGER: Spot Tests in Organic Analysis, 7th edn. (Engl. trans. by OESPER, R.E.), p. 185. Amsterdam: Elsevier, 1966.
399. SHERIF, E.A., R.K. GUPTA, and M. KRISHNAMURTI: Anomalous $AlCl_3$ Induced U.V. Shift of *C*-Alkylated Polyphenols. Tetrahedron Letters, **21**, 641 (1980).
400. JAIN, A.C., R.C. GUPTA, and P.D. SARPAL: Synthesis of (±) Lupinifolin, Di-*O*-Methyl Xanthohumol and Isoxanthohohumol and Related Compounds. Tetrahedron, **34**, 3563 (1978).

401. BOHLMANN, F., and C. ZDERO: Flavanones from *Helichrysum thapsus*. Phytochem., **22**, 2877 (1983).

402. FUKAI, T., and T. NOMURA: NMR Spectra of Isoprenoid Substituted Phenols. 6. ^{1}H NMR Spectra of Prenylated Flavonoids and Pyranoflavonoids. Heterocycles, **36**, 329 (1993).

403. ARNONE, A., G. CARDILLO, L. MERLINI, and R. MONDELLI: NMR Effects of Acetylation and Long-Range Coupling as a Tool for Structural Elucidation of Hydroxychromenes. Tetrahedron Letters, **1967**, 4201.

404. CHARI, V.M., S. AHMAD, and B.-G. ÖSTERDAHL: ^{13}C NMR Spectra of Chromeno- and Prenylated Flavones Structure Revision of Mulberrin, Mulberrochromene, Cyclomulberrin and Cyclomulberrochromene. Z. Naturforsch., **33b**, 1547 (1978).

405. SHIRATAKI, Y., M. KOMATSU, I. YOKOE, and A. MANAKA: Studies on the Constituents of *Sophora* Species. 16. Constituents of the Root of *Euchresta japonica* Hook. f. ex Regel. 1. Chem. Pharm. Bull. (Japan), **29**, 3033 (1981).

406. WEHRLI, F.W.: Proton-Coupled ^{13}C Nuclear Magnetic Resonance Spectra Involving ^{13}C-^{1}H Spin-Spin Coupling to Hydroxyl-Protons, a Complementary Assignment Aid. J. Chem. Soc., Chem. Commun. (London), **1975**, 663.

407. BRUNO, M., G. SAVONA, L. LAMARTINA, and F. LENTINI: New Flavonoids from *Bonannia graeca* (L.) Halacsy. Heterocycles, **23**, 1147 (1985).

408. SHIRATAKI, Y., I. YOKOE, M. ENDO, and M. KOMATSU: Determination of C-6 or C-8 Substituted Flavanone Using ^{13}C-^{1}H Long Range Coupling and the Revised Structures of Some Flavanones. Chem. Pharm. Bull. (Japan), **33**, 444 (1985).

409. FUKAI, T., and T. NOMURA: NMR Spectra of Isoprenoid Substituted Phenols. 3. Structure of 6- or 8-Isoprenoid Substituted Flavanone: Chemical Shift of the Hydrogen-Bonded Hydroxyl Group. Heterocycles, **31**, 1861 (1990).

410. FUKAI, T., and T. NOMURA: NMR Spectra of Isoprenoid Substituted Phenols. 9. Variations in the Chemical Shift of the 5-Hydroxyl Proton of 7-*O*-Prenylated Flavanones. Heterocycles, **43**, 1361 (1996).

411. WU, L.-J., T. MIYASE, A. UENO, M. KUROYANAGI, T. NORO, and S. FUKUSHIMA: Studies on the Constituents of *Sophora flavescens* Aiton. 2. Chem. Pharm. Bull. (Japan), **33**, 3231 (1985).

412. IINUMA, M., J. YOKOYAMA, M. OHYAMA, T. TANAKA, M. MIZUNO, and N. RUANGRUNGSI: Seven Phenolic Compounds in the Roots of *Sophora exigua*. Phytochem., **33**, 203 (1993).

413. BOHLMANN, F., C. ZDERO, H. ROBINSON, and R.M. KING: Naturally Occurring Terpene Derivatives. 357. A Diterpene, a Sesquiterpene Quinone and Flavanones from *Wyethia helenioides*. Phytochem., **20**, 2245 (1981).

414. FUKAI, T., and T. NOMURA: NMR Spectra of Isoprenoid Substituted Phenols. 4. Revised Structures of Albanins D and E, Geranylated Flavones from *Morus alba*. Heterocycles, **32**, 499 (1991).

415. TAKASUGI, M., S. ISHIKAWA, T. MASAMUNE, A. SHIRATA, and K. TAKAHASHI: Anti-Bacterial Compounds in the Branch Bark of Mulberry Tree. 42nd Annual Meeting of the Chemical Society of Japan, Abstract Paper, p. 352. Sendai, September 1980.

416a. FERRARI, F., I. MESSANA, and M. DO CARMO MESQUITA DE ARAUJO: Constituents of *Brosimopsis oblongifolia*. 2. Structures of Three New Flavones, Brosimones G, H, and I, form *Brosimopsis oblongifolia*. Planta Med., **55**, 70 (1989).

416b. FERRARI, F., and T. NOMURA: Letter to Editor. Planta Med., **58**, 116 (1992).

417. TAHARA, S., J.L. INGHAM, F. HANAWA, and J. MIZUTANI: ^{1}H NMR Chemical Shift Value of the Isoflavone 5-Hydroxyl Proton as a Convenient Indicator of 6-Substitution or 2′-Hydroxylation. Phytochem., **30**, 1683 (1991).

418. SCHRIPSEMA, J., and D. DAGNINO: Elucidation of the Substitution Pattern of 9,10-Anthraquinones Through the Chemical Shifts of *peri*-Hydroxyl Protons. Phytochem., **42**, 177 (1996).

419. ITO, C., T. MIZUNO, M. MATSUOKA, Y. KIMURA, K. SATO, I. KAJIURA, M. OMURA, M. JU-ICHI, and H. FURUKAWA: Constituents of Domestic *Citrus* Plants. 7. A New Flavonoid and Other New Components from *Citrus* Plants. Chem. Pharm. Bull. (Japan), **36**, 3292 (1988).

420. FUKAI, T., Y.-H. PEI, T. NOMURA, C.-Q. XU, L.-J. WU, and Y.-J. CHEN: Constituents of the Cultivated Mulberry Tree. 51. Phenolic Compounds from *Morus cathayana*. 2. 42nd Annual Meeting of Japanese Society of Pharmacognosy, Abstract Papers, p. 220. Fukuyama, September 1995.

421. FEENEY, J., and A. HEINRICH: The Use of Nuclear Magnetic Double Resonance to Detect and Characterise Phenolic OH Groups. J. Chem. Soc., Chem. Commun. (London), **1966**, 295.

422. GÜNTHER, H.: NMR Spectroscopy; Basic Principles, Concepts, and Applications in Chemistry 2nd edn., Chapter 4, Chichester, John Wiley & Sons, 1995.

423. VERBIT, L., and J.W. CLARK-LEWIS: Optically Active Aromatic Chromophores. 8. Studies in the Isoflavonoid and Rotenoid Series. Tetrahedron, **24**, 5519 (1968).

424. FUKAI, T., and T. NOMURA: Constituents of Licorice. 16. CD Spectra of Pyranoisoflavans. 41st Annual Meeting of Japanese Society of Pharmacognosy, Abstract Papers, p. 237. Sapporo, September 1994.

425. KUROSAWA, K., W.D. OLLIS, B.T. REDMAN, I.O. SUTHERLAND, H.M. ALVES, and O.R. GOTTLIEB: Absolute Configurations of Isoflavans. Phytochem., **17**, 1423 (1978).

426. YAHARA, S., T. OGATA, R. SAIJO, R. KONISHI, J. YAMAHARA, K. MIYAHARA, and T. NOHARA: Isoflavan and Related Compounds from *Dalbergia odorifera*. 1. Chem. Pharm. Bull. (Japan), **37**, 979 (1989).

427. HATANO, T., T. YASUHARA, K. MIYAMOTO, and T. OKUDA: Anti-Human Immunodeficiency Virus Phenolics from Licorice. Chem. Pharm. Bull. (Japan), **36**, 2286 (1988).

428. SHIBATA, S., H. INOUE, S. IWATA, R.-D. MA, L.-J. YU, H. UEYAMA, J. TAKAYASU, T. HASEGAWA, H. TOKUDA, A. NISHINO, H. NISHINO, and A. IWASHIMA: Inhibitory Effects of Licochalcone A Isolated from *Glycyrrhiza inflata* Root on Inflammatory Ear Edema and Tumour Promotion in Mice. Planta Med., **57**, 221 (1991).

429. CHEN, M., S. B. CHRISTENSEN, J. BLOM, E. LEMMICH, L. NADELMANN, K. FICH, T.G. THEANDER, and A. KHARAZMI: Licochalcone A, a Novel Anti-Parasitic Agent with Potent Activity against Human Pathogenic Protozoan Species of *Leishmania*. Antimicrob. Agents Chemother., **37**, 2550 (1993); Chem. Abstr., **120**, 68923e (1994).

430. KYOGOKU, K., K. HATAYAMA, S. YOKOMORI, R. SAZIKI, S. NAKANE, M. SASAJIMA, J. SAWADA, M. OHZEKI, and I. TANAKA: Anti-ulcer Effect of Isoprenyl Flavonoids. 2. Synthesis and Anti-ulcer Activity of New Chalcones related to Sophoradin. Chem. Pharm. Bull. (Japan), **27**, 2943 (1979).

431. KAN, Y.-M., Q. ZHU, L. CHEN, R.-D. WANG, X. LI, and M.-F. HONG: Effect of Homopterocarpin Separated from *Glycyrrhiza pallidiflora* on Human's Throat Cancer Cell (HEp-2). Zhongguo Yaoxue Zazhi (Chin. Pharm. Bull.), **29**, 608 (1994); Chem. Abstr., **122**, 23354v (1995).

432. MITSCHER, L.A., S. DRAKE, S.R. GOLLAPUDI, J.A. HARRIS, and D.M. SHANKEL: Isolation and Identification of Higher Plant Agents Active in Antimutagenic Assay Systems: *Glycyrrhiza glabra*. In: Basic Life Sci. 39: Antimutagenenesis and Anticarcinogenesis Mechanisms. (SHANKEL, D.M., P.E. HARTMAN, T. KADA, and

A. Hollaender, eds.), pp. 153–165, New York: Plenum Press, 1986; Chem. Abstr., **106**, 170739x (1987).

433. Kimura, Y., H. Okuda, T. Okuda, and S. Arichi: Effects of Chalcones Isolated from Licorice Roots on Leukotriene Biosynthesis in Human Polymorphonuclear Neutrophils. Phytother. Res., **2**, 140 (1988); Chem. Abstr., **110**, 33463w (1989).

434. Okada, K, Y. Tamura, M. Yamamoto, Y. Inoue, R. Takagaki, K. Takahashi, S. Demizu, K. Kajiyama, Y. Hiraga, and T. Kinoshita: Identification of Antimicrobial and Antioxidant Constituents from Licorice of Russian and Xinjiang Origin. Chem. Pharm. Bull. (Japan), **37**, 2528 (1989).

435. Vakhabov, A.A., C.D. Aminov, and R.K. Hasanova: Antiinflammatory Activity of Pinocembrin. Dokl. Akad. Nauk Resp. Uzb. (USSR), **1993**, 43; Chem. Abstr., **122**, 96053f (1995).

436. Kobayashi, S., T. Miyamoto, I. Kimura, and M. Kimura: Inhibitory Effect of Isoliquiritin, a Compound in Licorice Root, on Angiogenesis *in Vivo* and Tube Formation *in Vitro*. Biol. Pharm. Bull. (Japan), **18**, 1382 (1995).

437. Yamamoto, K., H. Kakegawa, H. Ueda, H. Matsumoto, T. Sudo, T. Miki, and T. Satoh: Gastric Cytoprotective Anti-Ulcerogenic Actions of Hydroxychalcones in Rats. Planta Med., **58**, 389 (1992).

438. Tawata, M., K. Aida, T. Noguchi, Y. Ozaki, S. Kume, H. Sasaki, M. Chin, and T. Onaya: Anti-Platelet Action of Isoliquiritigenin, an Aldose Reductase Inhibitor in Licorice. Eu. J. Pharmac., **212**, 87 (1992).

439. Tawata, M., Y. Yoda, K. Aida, H. Shindo, H. Sasaki, M. Chin, and T. Onaya: Anti-Platelet Action of GU-7, a 3-Arylcoumarin Derivative, Purified from Glycyrrhizae Radix. Planta Med., **56**, 259 (1990).

440. Huang, L.-Y., and Y.K.T. Lam: Isolation of 3′,6′-Diisopentenyl-2′,4′-dihydroxy-5,7-dimethoxyisoflavan and Compositions Containing It and Other Related Compounds as Anxielytic Agents and Avermectin Antidotes. US patent; Chem. Abstr., **106**, 189006c (1987).

441. Tanaka, S., Y. Kuwai, and M. Tabata: Isolation of Monoamine Oxidase Inhibitors from *Glycyrrhiza uralensis* Roots and the Structure-Activity Relationship. Planta Med., **53**, 5 (1987).

442a. Aida, K., M. Tawata, H. Shindo, T. Onaya, H. Sasaki, H. Nishimura, M. Chin (Z. Chen), H. Mitsuhashi: The Existence of Aldose Reductase Inhibitors in Some Kampo Medicines (Oriental Herb Prescriptions). Planta Med., **55**, 22 (1989).

442b. Aida, K., M. Tawata, H, Shindo, T. Onaya, H. Sasaki, T. Yamaguchi, M. Chin (Z. Chen), and H. Mitsuhashi: Isoliquiritigenin: a New Aldose Reductase Inhibitor from Glycyrrhizae Radix. Planta Med., **56**, 254 (1990).

442c. Benvenuti, S., F. Severi, L. Costantino, M. Melegari, and A. Mucci: Identification, Characterization, and Biological Activity of Chalcone Derivatives of *Glycyrrhiza glabra* L. Riv. Ital. EPPOS, **7**, 13 (1996); Chem. Abstr., **126**, 222831 (1997).

443. Kusano, A., T. Nikaido, T. Kuge, T. Ohmoto, G.D. Monache, B. Botta, M. Botta, and T. Saitoh: Inhibition of Adenosine 3′,5′-Cyclic Monophosphate Phosphodiesterase by Flavonoids from Licorice Roots and 4-Arylcoumarins. Chem. Pharm. Bull. (Japan), **39**, 930 (1991).

444. Kakegawa, H., H. Matsumoto, and T. Satoh: Inhibitory Effects of Some Natural Products on the Activation of Hyaluronidase and Their Antiallergic Actions. Chem. Pharm. Bull. (Japan), **40**, 1439 (1992).

445a. Sato, S., T. Yanagisawa, H. Mitsuhashi, and T. Nomura: Allergy Inhibitors Containing Flavonoids from Mulberry, Licorice, or *Epimedium*. Jpn. patent; Chem. Abstr., **115**, 231980d (1991).

445b. SATO, S., M. CHIN, H. MITSUHASHI, and T. NOMURA: Sodium-potassium-activated ATPase Inhibitors Containing Flavonoids from Mulberry or Licorice. Jpn. patent; Chem. Abstr., **115**, 182941f (1991).

446. YAMAGUCHI, T., S. SATO, H. MITSUHASHI, T. NOMURA: Aldose Reductase Inhibitors Containing Benzopyrans for Treatment of Diseases associated with Diabetes. Jpn. patent; Chem. Abstr., **116**, 46283m (1992).

447. FUKAI, T., L. ZENG, J. NISHIZAWA, and T. NOMURA: Studies of Prenylated Flavonoids of Chinese Licorices. 9th Symposium on the Development and Application of Naturally Occurring Drug Material, Abstract of Papers, p. 49. Shizuoka (Japan), July 1993.

448. YANG, S.-L., and Y.-L. LIU: Chemical Constituents of *Glycyrrhiza inflata* Bat. Zhiwu Xuebao (Acta. Bot. Sinica, China), **30**, 176 (1988); Chem. Abstr., **109**, 107754v (1988).

449. ELGAMAL, M.H.A., F.K. ABDEL HADY, A.G. HANNA, G.H. MAHRAN, and H. DUDDECK: A further Contribution to the Triterpenoid Constituents of *Glycyrrhiza glabra* L. Z. Naturforsch., **45c**, 937 (1990).

450. TSUBONE, K., S. OHNISHI, and T. YONEYA: Separation of Glycyrrhizinic Acid Isomers by High-Performance Liquid Chromatography. J. Chromatogr., **248**, 469 (1982).

451. ZHAO, Y.-Y., R.-Y. ZHANG, and M. LIU: Chemical Constituents of *Glycyrrhiza inflata* Bat. 1. Beijing Yike Daxue Xuebao (J. Beijing Med. Univ., China), **22**, 283 (1990); Chem. Abstr., **117**, 44632w (1992).

452. PEDERIVA, R., and O.S. GIORDANO: Flavonoids from *Glycyrrhiza astragalina*. An. Asoc. Quim. Argent, **75**, 7 (1987); Chem. Abstr., **107**, 112718z (1987).

453. GAO, D.-Y., and R.-Y. ZHANG: Chemical Constituents of Malay Licorice (*Glycyrrhiza yunnanensis*). Zhongcaoyao (Chin. Herbal Med.), **25**, 507 (1994); Chem. Abstr., **122**, 89199e (1995).

454. KOBAYASHI, M., H. NOGUCHI, and U. SANKAWA: Formation of Chalcones and Isoflavones by Callus Culture of *Glycyrrhiza uralensis* with Different Production Patterns. Chem. Pharm. Bull. (Japan), **33**, 3811 (1985).

455. LIU, J.-H., S.-S. YANG, Y.-Q. FU, D. XU, C.-X. JIANG, and F.-F. HOU: Flavonoid Contents of *Glycyrrhiza pallidiflora*. Zhongcaoyao (Chin. Herbal Med.), **23**, 349 (1992); Chem. Abstr., **117**, 230157v (1992).

456. LIANG, H., and R.-Y. ZHANG: Chemical Constituents of *Glycyrrhiza squamulosa*. Beijing Yike Daxue Xuebao (J. Beijing Med. Univ., China), **24**, 399 (1992); Chem. Abstr., **118**, 209446z (1993).

457. ZOU, K., F. DI, R.-Y. ZHANG, L. ZENG, and N.-W. FU: Constituents of *Glycyrrhiza inflata* Bat. 3. Tianran Chanwu Yanjiu Yu Kaifa (Nat. Prod. Research Develop., China), **5** (4), 1 (1993); Chem. Abstr., **120**, 319343n (1994).

458. HOU, S.-S., S.-Y. CHEN, M.-Z. YANG, X.-M. LI, D.-Y. LU, J. ZHANG, and L. CHEN: Isolation and Identification of Echinatin from Cultured Cells of *Glycyrrhiza uralensis*. Zhowu Xuebao (Acta Bot. Sinica China), **35**, 567 (1993); Chem. Abstr., **120**, 73474p (1994).

459. KATTAEV, N.S., and G.K. NIKONOV: Flavonoids of *Glycyrrhiza glabra*. Khim. Prir. Soedin., **10**, 93 (1974); Chem. Abstr., **81**, 60809c (1974).

460. HAYASHI, H., M. YASUMA, N. HIRAOKA, Y. IKESHIRO, H. YAMAMOTO, E. YESILADA, E. SEZIK, G. HONDA, and M. TABATA: Field Survey of Licorice in Turkey. 3. Flavonoids Variation in the Leaves of *Glycyrrhiza glabra*. Phytochem., **42**, 701 (1996).

461. BATIROV, E. K., F. KIYAMITDINOVA, and V.M. MALIKOV: Flavonoids of the Aerial Parts of *Glycyrrhiza glabra*. Khim. Prir. Soedin., **1986**, 111; Chem. Abstr., **104**, 203930s (1986).

462. Shibano, M., Y. Matsumoto, G. Kusano, and T. Shibata: Researches of *Glycyrrhiza* Species Grown at Medicinal Plant Gardens in Japan and Basic Studies for Selection of Pharmaceutically Fine Races. 1. Comparative Studies by HPLC Patterns and Constituents of Aerial parts. Nat. Med. (Jpn. J. Pharmacognosy), **50**, 273 (1996); Chem. Abstr., **125**, 338818 (1996).

463. Hu. J.-F., Z.-L. Ye, and F.-G. Shen: Constituents from the Roots of *Glycyrrhiza yunnanensis*. 1. Tianran Chanwu Yanjiu Yu Kaifa (Nat. Prod. Research Develop., China), **6** (4), 6 (1994); Chem. Abstr., **123**, 5600p (1995).

464. Wang, C.-L., R.-Y. Zhang, Y.-S. Han, X.-G. Dong, and W.-B. Liu: Chemical Studies of Coumarins from *Glycyrrhiza uralensis* Fisch. Yaoxue Xuebao (Acta Pharm. Sinica, China), **26**, 147 (1991); Chem. Abstr., **115**, 68400d (1991).

465. Litvinenko, V.I., N.P. Maksyutina, and D.G. Kolesnikov: Flavonoidal Compounds of *Glycyrrhiza glabra*. 1. Flavonoid L-1. Zhurn. Obshchei Khimii (USSR), **33**, 4014 (1963); Chem. Abstr., **60**, 9347 (1964).

466. Afchar, D., A. Cavé, H. Guinaudeau, and J. Vaquette: Study of Licorices from Iran. 3. Flavonoids from *Glycyrrhiza echinata* L. Roots. Plant. Med. Phytother., **18**, 170 (1984); Chem. Abstr., **102**, 128851v (1985).

467. Afchar. D., A. Cavé, and J. Vaquette: Study on Licorice from Iran. 2. Flavonoids from *Glycyrrhiza glabra* L. var. *glandulifera* Waldst. *et* Kit. and *Glycyrrhiza glabra* L. var. *violacea* Boiss. Plant. Med. Phytother., **18**, 55 (1984).

468. Litvinenko, V.I., and T.P. Nadezhina: Flavonoids of the Aerial Part of *Glycyrrhiza macedonica*. Rast. Resur., **6**, 575 (1970); Chem. Abstr., **74**, 95410y (1971).

469. Elgamal, M.H.A., and M.B.E. Fayez: Constituents of Local Plants. 16. Isolation of Formononetin from the Roots of *Glycyrrhiza glabra* Linn. Collected Locally. Indian J. Chem., **10**, 128 (1972).

470. Arias-Castro, C., A.H. Scragg, A. Stafford, and M. Rodriguez-Mendiola: Isolation, Identification, and Kinetics of Production of Formononetin in Cell Suspension Cultures of *Glycyrrhiza glabra*. Planta Med., **56**, 631 (1990).

471. Hayashi, H., G. Honda, M. Tabata, H. Yamamoto, E. Yesilada, and E. Sezik: Field Survey of Licorice in Turkey. 2. A Survey of Distribution and Characteristics of *Glycyrrhiza glabra* L. in Turkey. Nat. Med. (Jpn. J. Pharmacognosy), **49**, 129 (1995); Biol. Abstr., **100**, 98295 (1995).

472. Saleh, M.M., M.M. El-Olemy, H.M. Metawie, A.H. Abou-Zied, and M.S. Abdel-Salam: Response of Liquorice (*Glycyrrhiza glabra*) Leaves to Certain Stress Factor. Planta Med., **56**, 610 (1990).

473. Reiners, W.: Components of Licorice Roots. 2. 7-Hydroxy-4′-methoxyisoflavone (Formononetin) from Licorice Root. Experientia, **22**, 359 (1966).

474. Balbaa, S.I., G.H. Mahran, G.A. El-Hossary, and M.A. Selim: A Phytochemical Study of *Glycyrrhiza glabra* L. Growing in Egypt. Bull. Fac. Pharm., Cairo Univ., **14**, 213 (1976); Chem. Abstr., **87**, 206410p (1977).

475. Oda, M., Y. Tamura, K. Mizutani, and O. Tanaka: Biological Activity and Constituents of Cell Cultures of Licorice. 42nd Annual Meeting of Japanese Society of Pharmacognosy, Abstract Papers, p. 168. Fukuyama, September 1995.

476. Raggi, M.A., F. Bugamelli, L. Nobile, V. Curcelli, R. Mandrioli, A. Rossetti, and G. Cantelli forti: Studies on the Choleretic Effect of Licorice: Identification and Determination of Pharmacologically Active Components of *Glycyrrhiza glabra*. Boll. Chim. Farm., **134**, 634 (1995); Chem. Abstr., **124**, 278675 (1996).

477. Zou, K., Y.-I. Zhao, N.-W. Fu, L. Qiao, and R.-Y. Zhang: Antioxidant Constituents from *Glycyrrhiza inflata* Bat. Root. J. Chin. Pharm. Sci., **5**, 182 (1996); Chem. Abstr., **126**, 273418 (1997).

478. ASADA, Y., W. LI, and T. YOSHIKAWA: Constituents of Hairy Root Cultures of Licorice. 117th Annual Meeting of Pharmaceutical Society of Japan, Abstract Papers, part 2, p. 117. Tokyo, March 1997.

479. TAKAGI, M., H. ITOH, T. HATANO, and T. YOSHIDA: Polyphenols of Licorice. 14. 117th Annual Meeting of Pharmaceutical Society of Japan, Abstract Papers, part 2, p. 132. Tokyo, March 1997.

(*Received January 3, 1997*)

Author Index

Page numbers printed in *italics* refer to References

Subject Index

Springer-Verlag and the Environment

WE AT SPRINGER-VERLAG FIRMLY BELIEVE THAT AN international science publisher has a special obligation to the environment, and our corporate policies consistently reflect this conviction.

WE ALSO EXPECT OUR BUSINESS PARTNERS – PRINTERS, paper mills, packaging manufacturers, etc. – to commit themselves to using environmentally friendly materials and production processes.

THE PAPER IN THIS BOOK IS MADE FROM NO-CHLORINE pulp and is acid free, in conformance with international standards for paper permanency.

MIX
Papier aus verantwortungsvollen Quellen
Paper from responsible sources
FSC® C105338

If you have any concerns about our products,
you can contact us on
ProductSafety@springernature.com

In case Publisher is established outside the EU,
the EU authorized representative is:
Springer Nature Customer Service Center GmbH
Europaplatz 3, 69115 Heidelberg, Germany

Printed by Libri Plureos GmbH
in Hamburg, Germany